Gear Noise and Vibration

MECHANICAL ENGINEERING
A Series of Textbooks and Reference Books

Founding Editor

L. L. Faulkner

*Columbus Division, Battelle Memorial Institute
and Department of Mechanical Engineering
The Ohio State University
Columbus, Ohio*

Additional Volumes in Preparation

Gear Noise and Vibration

Second Edition, Revised and Expanded

J. Derek Smith
Cambridge University
Cambridge, England

MARCEL DEKKER, INC. NEW YORK • BASEL

FIRST INDIAN REPRINT 2005

Library of Congress Cataloging-in-Publication Data
A catalog record for this book is available from the Library of Congress.

ISBN : 0-8247-4129-3

Headquarters
Marcel Dekker, Inc.
270 Madison Avenue, New York, NY 10016
tel: 212-696-9000; fax: 212-685-4540

Eastern Hemisphere Distribution
Marcel Dekker AG
Hutgasse 4, Postfach 812, CH-4001 Basel, Switzerland
tel: 41-61-260-6300; fax: 41-61-260-6333

World Wide Web
http://www.dekker.com

The publisher offers discounts on this book when ordered in bulk quantities. For
more information, write to Special Sales/Professional Marketing at the headquarters
address above.

Current printing (last digit):
10 9 8 7 6 5 4 3 2 1

Printed in India, Brijbasi Art Press Ltd., I-72, Sector-9, Noida, U.P. India.
FOR SALE IN THE INDIAN SUBCONTINENT ONLY

To Rona

Preface to the Second Edition

Since the first edition there have been many changes in the equipment available for measurements and the growing interest in Transmission Error measurement has spawned numerous approaches that are not always clearly described. Each author has a tendency to extoll the virtues of his approach but rarely points out the corresponding disadvantages, so I have attempted to compare systems. A range of new problems in from industry has generated some interesting additional topics.

I have also added discussion of some of the less common but puzzling topics such as high contact ratio gears which are increasingly being used to reduce noise. Testing procedures are also discussed in more detail together with some practical problems and some slightly extended description of the failures that may be encountered and their relationship, or lack of it, to noise problems.

I hope that few errors or mistakes have crept into the book but if readers discover errors I will be very grateful if they let me know (e-mail jds1002@eng.cam.ac.uk)

J. Derek Smith

Preface to the First Edition

This discussion of gear noise is based on the experience of nearly 40 years of researching, consulting, measuring and teaching in the field of gears, mainly biased towards solving industrial noise and vibration problems.

When a noise or vibration problem arises there is usually a naive hope either that it will go away or that slapping on a layer of Messrs. Bloggs' patent goo will solve the problem. Unfortunately, gear problems are hidden beneath the skin so they cannot normally be cured simply by treating the symptoms and they rarely disappear spontaneously. Another hope is that by going to an "expert" who has a very large, sophisticated (expensive) software program there will be a simple solution available without the boring need to find out exactly what is causing the trouble at the moment.

Neither approach is very productive. In addition, anything to do with gears is unpopular because of the strange jargon of gears, especially where "corrections" are involved and the whole business is deemed to be a rather "black art." Those few who have mastered the "black art" tend to be biased towards the (static) stressing aspects or the manufacturing of gears. So they recoil in horror from vibration aspects since they involve strange ideas such as electronics and Fast Fourier Transforms. In practice few "experts" will get down to the basics of a problem since understanding is often lacking and measurements may not be possible. Vibration "experts" tend to be so concerned with the complex, elegant mathematics of some esoteric analysis techniques that they are not interested in basic causes and explanations.

Gear books have traditionally concentrated on the academic geometry of gears (with "corrections") and have tended to avoid the difficult, messy, real engineering of stresses and vibrations. The area of stresses is well covered by the various official specifications such as DIN 3990 and the derived ISO 6336 and BS 436 and the rival AGMA 2001, all based on a combination of (dodgy) theory and practical testing. Since it is usually necessary for the manufacturer to keep to one of the specifications for legal reasons, there is no point in departing from the standard specifications. In the area of noise and vibration, my previous book (Marcel Dekker, 1983) was written rather a long time ago and the subject has moved on greatly since then. Prof. Houser gives a good summary of gear noise in a chapter in the 1992 version of *Dudley's Gear Handbook* (McGraw-Hill) with many references.

This book is intended to help with the problems of design, metrology, development and troubleshooting when noise and vibration occur. In this area the standard specifications are of no help, so it is necessary to understand what is happening to cause the noise. It is intended primarily for engineers in industry who are either buying-in gears or designing, manufacturing, and inspecting them and who encounter noise trouble or are asked to measure strange, unknown quantities such as Transmission Error (T.E.). It should also be of interest to graduate students or those in research who wish to understand more about the realities of gears as part of more complex designs, or who are attempting to carry out experiments involving gears and are finding that dynamics cannot be ignored.

I have attempted to show that the design philosophy, the geometry, and the measurement and processing of the vibration information are relatively straightforward. However, any problem needs to be tackled in a reasonably logical manner, so I have concentrated on basic non-mathematical ideas of how the vibration is generated by the T.E. and then progresses through the system. Mathematics or detailed knowledge of computation are not needed since it is the understanding, the measurement, and the subsequent deductions that are important. It is measurement of reality that dominates the solution of gear problems, not predictions from software packages. It is also of major importance to identify whether the problems arise from the gears or from the installation, and this is best done experimentally.

I hope that this book will help researchers and development engineers to understand the problems that they encounter and to tackle them in an organised manner so that decisions to solve problems can be taken rationally and logically.

This book owes much to many friends, colleagues, and helpers in academia and in industry who have taught me and broadened my knowledge while providing many fascinating problems for solution.

J. Derek Smith

Contents

Contents   **xi**

1

Causes of Noise

1.1 Possible causes of gear noise

To generate noise from gears the primary cause must be a force variation which generates a vibration (in the components), which is then transmitted to the surrounding structure. It is only when the vibration excites external panels that airborne noise is produced. Inside a normal sealed gearbox there are high noise levels but this does not usually matter since the air pressure fluctuations are not powerful enough to excite the gearcase significantly. Occasionally in equipment such as knitting machinery there are gears which are not sealed in oiltight cases and direct generated noise can then be a major problem.

There are slight problems in terminology because a given oscillation at, for example, 600 Hz is called a vibration while it is still inside the steel but is called noise as soon as it reaches the air. Vibrations can be thought of as either variations of force or of movement, though, in reality, both must occur together. Also, unfortunately, mechanical and electrical engineers often talk about "noise" when they mean the background random vibrations or voltages which are not the signal of interest. Thus we can sometimes encounter something being described as the signal-to-noise ratio of the (audible) noise. An additional complication can arise with very large structures especially at high frequencies because force and displacement variations no longer behave as conventional vibrations but act more as shock or pressure waves radiating through the system but this type of problem is rare.

In general it is possible to reduce gear noise by:

(a) Reducing the excitation at the gear teeth. Normally for any system, less amplitude of input gives less output (noise) though this is not necessarily true for some non-linear systems.

(b) Reducing the dynamic transmission of vibration from the gear teeth to the sound radiating panels and out of the panels often by inserting vibration isolators in the path or by altering the sound radiation properties of the external panels.

(c) Absorbing the noise after it has been generated or enclosing the whole system in a soundproof box.

(d) Using anti-noise to cancel the noise in a particular position or limited
    number of positions, or using cancellation methods to increase the
    effectiveness of vibration isolators.

Of these approaches, (c) and (d) are very expensive and tend to be
temperamental and delicate or impracticable so this book concentrates on (a)
and (b) as the important approaches, from the economic viewpoint.
Sometimes initial development work has been done by development engineers
on the gear resonant frequencies or the gear casing or sound radiating
structure so (b) may have been tackled in part, leaving (a) as the prime target.
However, it is most important to determine first whether (a) or (b) is the
major cause of trouble.

A possible alternative cause of noise in a spur gearbox can occur
with an overgenerous oil supply if oil is trapped in the roots of the meshing
teeth.   If the oil cannot escape fast through the backlash gap, it will be
expelled forcibly axially from the tooth roots and, at once-per-tooth
frequency, can impact on the end walls of the gearcase.   This effect is rare
and does not occur with helical teeth or with mist lubrication.

The excitation is generally due to a force varying either in
amplitude, direction or position as indicated in Fig. 1.1. Wildhaber-Novikov
or Circ-Arc gears [1] produce a strong vibration excitation due to the
resultant force varying in position [Fig. 1(c)] as the contact areas move
axially along the pitch line of the gears, so this type of drive is inherently
noisier than an involute design.

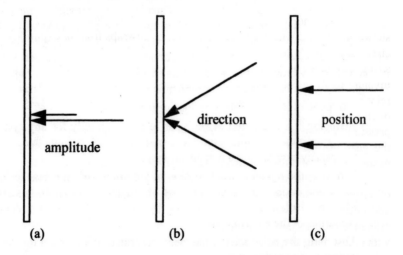

Fig 1.1   Types of vibration excitation due to change in amplitude (a),
direction (b), or position (c).

Variation of direction of the contact force between the gears [Fig. 1(b)] can occur with unusual gear designs such as cycloidal and hypocycloidal gears [2] but, with involute gears, the direction variation is only due to friction effects. The effect is small and can be neglected for normal industrial gears as it is at worst a variation of ± 3° when the coefficient of friction is 0.05 with spur gears but is negligibly small with helical gears.

For involute gears of normal attainable accuracy it is variation of the amplitude of the contact force [Fig..1(a)] that gives the dominant vibration excitation. The inherent properties of the involute give a constant force direction and a tolerance of centre distance variation as well as, in theory, a constant velocity ratio.

The source of the force variation in involute gears is a variation in the smoothness of the drive and is due to a combination of small variations of the form of the tooth from a true involute and varying elastic deflection of the teeth. This relative variation in displacement between the gears acts via the system dynamic response to give a force variation and resulting vibration.

This book deals mainly with parallel shaft involute gears since this type of drive dominates the field of power transmission. Fundamentally the same ideas apply in the other types of drive such as chains, toothed belts, bevels, hypoids, or worm and wheel drives but they are of much less economic importance. The approach to problems is the same.

1.2 The basic idea of transmission error

The fundamental concept of operation of involute (spur) gears is that shown in Fig. 1.2 where an imaginary string unwraps from one (pinion) base circle and reels onto a second (wheel) base circle. Any point fixed on the string generates an involute relative to base circle 1 and so maps out an involute tooth profile on gear 1 and at the same time maps out an involute relative to gear 2. (An involute is defined as the path mapped out by the end of an unwrapping string.) This theoretical string is the "line of action" or the pressure line and gives the direction and position of the normal force between the gear teeth. Of course it is a rather peculiar mathematical string that pushes instead of pulls, but this does not affect the geometry.

In the literature on gearing geometry there is a tremendous amount of jargon with much discussion of pitch diameters, reference diameters, addendum size, dedendum size, positive and negative corrections (of the reference radius), undercutting limits, pressure angle variation, etc., together with a host of arcane rules about what can or cannot be done.

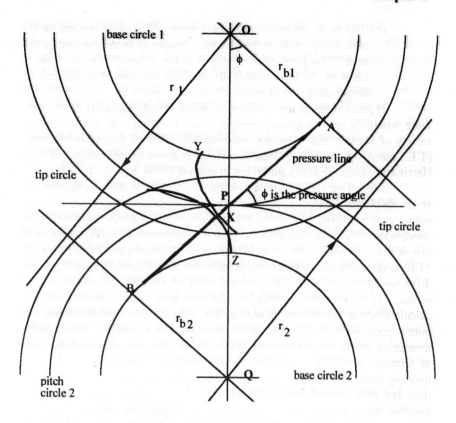

Fig 1.2 Involute operation modelled on unwrapping string.

All this is irrelevant as far as noise is concerned and it is important to remember that the involute is very, very simply defined and much jargon merely specifies where on an involute we work.

There is, in reality, only one true dimension on a spur gear and that is the base circle radius (and the number of teeth). Any one involute should mate with another to give a constant velocity ratio while they are in contact. It is possible to have two gears of slightly different nominal pressure angle meshing satisfactorily since pressure angle is not a fundamental property of a flank and depends on the centre distance at which the gears happen to be set. The only relevant criteria are:

(a) Both gears must be (nearly) involutes.

(b) Before one pair of teeth finish their contact the next pair must be ready to take over (contact ratio greater than 1.00).

(c) The base pitches of both gears must be the same (except for tip relief) so that there is a smooth handover from one pair to the next. (The base pitch of a gear is the distance from one tooth's flank to the next tooth's flank along the line of action and so tangential to the base circle.)

If gears were perfect involutes, absolutely rigid and correctly spaced, there would be no vibration generated when meshing. In practice, for a variety of reasons, this does not occur and the idea of Transmission Error (T.E.) came into existence. Classic work on this was carried out by Gregory, Harris and Munro [3,4] at Cambridge in the late 1950s.

We define T.E. [5] by imagining that the input gear is being driven at an absolutely steady angular velocity and we would then hope that the output gear was rotating at a steady angular velocity. Any variation from this steady velocity gives a variation from the "correct position" of the output and this is the T.E. which will subsequently generate vibration. More formally, "T.E. is the difference between the angular position that the output shaft of a drive would occupy if the drive were perfect and the actual position of the output." In practical terms, we take successive angular positions of the input, calculate where the output should be, and subtract this from the measured output position to give the "error" in position. Measurements are made by measuring angular displacements and so the answers appear initially in units of seconds of arc. It is possible to measure T.E. semi-statically by using dividing heads and theodolites on input and output and indexing a degree at a time but this is extremely slow and laborious though it can be the only possible way for some very large gears. Although the measurements are made as angular movements the errors are rarely given as angles as it is much more informative to multiply the error angle (in radians) by the pitch circle radius to turn the error into microns of displacement. Such errors are rather small typically only a micron or two even for mass produced gears such as those in cars.

There is, unfortunately, some uncertainty as to whether we should multiply by pitch circle radius to get tangential movement at pitch radius or multiply by base circle radius to get movement along the pressure line, i.e., normal to the involute surfaces. Either is legitimate but we usually use the former since it ties in with the standard way of defining pitch and helix errors between teeth. However, from a geometric aspect, to correspond with profile error measurements (which are normal to the involute), the latter is preferable.

The great advantage of specifying T.E. as a linear measurement (typically less than 5 μm) is that all gears of a given quality, regardless of size of tooth module or pitch diameter, have about the same sizes of error so comparisons are relatively easy.

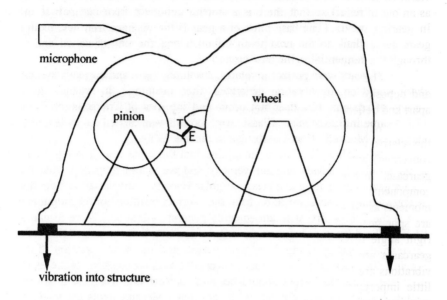

Fig 1.3 Transmission error excitation between gears.

It seems utterly ridiculous that a 1 mm module (25DP) gear less than an inch diameter will have roughly the same T.E. as a 25 mm module (1DP) wheel some 3 metres diameter of the same quality, but this is surprisingly close to what happens in practice (the module is the pitch circle diameter of the gear in millimetres divided by the number of teeth).   This unexpected constant size of errors is liable to cause problems in the future with the current trend towards "micromechanics". If a gear tooth is only 20 μm tall, the base pitch is about 20 μm but errors of 2 μm in pitch or profile are still likely with corresponding T.E. errors so that a speed variation of 10% becomes possible.

Having defined T.E., we are left with a mental picture either of the "unwrapping string" varying in length or, as sketched in Fig. 1.3, of a small but energetic demon between the gear teeth surfaces imposing a relative vibration. For most noise purposes it is only the vibrating part of the T.E. that is important so any steady (elastic) deflections are ignored.

1.3 Gearbox internal responses

T.E. is the error between the gear teeth.   This idea of a relative displacement (microns) being the cause of a force variation and hence vibration is unusual since traditionally we excite with an external force such as an out of balance or vibrate the supporting ground to produce a vibration. In gearing we have a relative displacement (the T.E.) between the mating gears generating the forces between the teeth and the subsequent vibrations through the system.

The relative displacement between the teeth is generated by equal and opposite vibrating forces on the two gear teeth surfaces, moving them apart and deflecting them a sufficient distance to accommodate the T.E.

When we consider the internal responses of the gearbox, the input is the relative vibration between the gear teeth and the outputs (as far as noise is concerned) are the vibration forces transmitted through the bearings to the gearcase. In general the "output" force through each bearing should have six components: three forces and three moments, but we usually ignore the moments as they are very small and the axial forces will be negligible if there are spur gears, double helicals, or thrust cones. Single helical gears (and right angle drives) give axial forces and, unfortunately, the end panels of gearcases are often flat and are rather flexible. The resulting end panel vibrations are important if it is the gearcase which is producing noise, but of little importance if it is vibration through the mounting feet that is the principal cause.

Occasionally vibrating forces will transmit along the shafts to outside components and radiate noise. A ship's propeller will act as a good loudspeaker if directly coupled to a gearbox, but insertion of a flexible elastomeric coupling will usually block the vibration effectively, provided it has been correctly designed for the right frequency range. Similarly, in wind turbines, the propellors can act as surprisingly effective loudspeakers so it is necessary to have good isolation between blades and gears. In a car, the trouble path can be upstream or downstream, as vibration from the gearbox travels to the engine and radiates from engine panels, or escapes through the engine mounts to the body shell, or travels to the rear axle and through its supports to the body. At one time the vibration also travelled directly via gear levers and clutch cables into the body shell.

The assumption usually made is that, when modelling internal resonances and responses, the bearing housings can be taken as rigid. This is usually a reasonable idealisation of the situation since bearing housing movements are typically less than 10% of gear movements. Occasionally a flexible casing, or one where masses are moving in antiphase, will give the effect of reducing or increasing the apparent stiffness of supporting shafts or bearings.

Gears are sometimes assumed to vibrate only torsionally but this assumption is wildly incorrect due to bearings and to shaft deflections so any model of gears must allow for lateral movement (i.e., movement perpendicular to the gear axis). Masses are known accurately and stiffnesses can be predicted or measured with reasonable precision, but there are major problems with damping which cannot be designed or predicted reliably.

1.4 External responses

The path of the vibration from the bearing housings to the final radiating panels on either the gearcase or external structure is usually complex. Fortunately, although prediction is difficult and unreliable due to damping uncertainties it is relatively easy to test experimentally so this part of the path rarely gives much trouble in development.

One of the first requirements is to establish whether it is the gearcase itself which is the dominant noise source or, more commonly, whether the vibration is transmitted into the main structure to generate the noise. Transmission to the structure is greatly affected by the isolators fitted between the gearbox and the structure.

There is liable to be a large number of parallel paths for the vibration through the structure and an extremely large number of resonances which are so closely packed in frequency that they overlap. A statistical energy approach [6] with the emphasis on energy transmission and losses over a broad frequency band can give a clearer description than the conventional transfer function approach when frequencies are high and there are multiple inputs and resonances. In a very large structure the conventional ideas of resonant systems are no longer so relevant and the transmission of energy has more in common with ideas of propagation of stress waves.

1.5 Overall path to noise

The complete vibration transmission path is shown in Fig. 1.4. It starts from the combination of manufacturing errors, design errors and tooth and gear deflections to generate the T.E. Though manufacturing errors are usually blamed it is more commonly design that is at fault.

The T.E. is then the source of the vibration and it drives the internal dynamics of the gears to give vibration forces through the bearing supports. In turn, these bearing forces drive the external gearcase vibrations or, via any isolation mounts, drive the external structure to find "loudspeaker" panels. In a vehicle, after the vibration has travelled from the gearbox through the engine main casting to the support mounts and hence to the structure, it may travel several metres in the body before exciting a panel to emit sound that annoys the occupants. Vibration travelling along te input and output shafts to cause trouble can aalso occur but is less common.

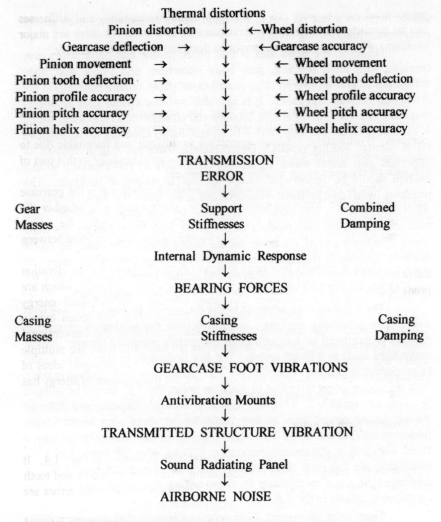

Fig 1.4 Vibration excitation and transmission path.

1.6 T.E. - noise relationship

It is very difficult for a traditional gear engineer trained to think in terms of pitch, profile, and helix measurements to change over to ideas of single flank checking, i.e., T.E., especially as T.E. is not relevant for gear strength. The change is not helped by the difference that the traditional methods are methods where the gears are stationary on expensive machines in the metrology lab whereas T.E. is measured when the gears are rotating and can be done on a test rig out in the main works or sometimes even on the equipment while running normally.

However, the basic idea is that pitch, profile and helix errors may combine with tooth bending, gear body distortions and whole gear body deflections to give an overall relative deflection (from smooth running) at the meshpoint between the gears. It is also difficult to convince gear engineers that there is a very big difference between roll (double flank) checking, which is extremely cheap and easy, and T.E. (single flank) checking since they give rather similar looking results. Unfortunately, there are a large number of important gear errors which are missed completely by roll checking so this method should be discouraged except for routine control of backlash. The problems with double flank measurement arise from the basic averaging effect that occurs. Any production process or axis error in transfer from machine to machine may produce errors which give +ve errors on one flank which effectively cancel -ve errors on the facing flank. The resulting centre distance variation is negligible but there may be large (cancelling) errors on the drive and overrun flanks. Shavers and certain types of gear grinders are prone to this type of fault which is worse with high helix angle gears.

The question then arises as to the connection between T.E. and final noise. Few practising engineers initially believe the academics' claim that noise is proportional to T.E., although the system normally behaves (except under light load) as a linear system. For any linear system the output should be proportional to input. Doubling the T.E. should give 6dB increase in noise level or, with a target reduction of 10dB on noise, the T.E. should be reduced by $\sqrt{10}$, i.e., roughly 3. This only applies at a single frequency and different frequencies encounter high or low responses en route so a major visible frequency component in the T.E. may be minor in the final noise because it could not find a convenient resonance. Tests over 20 years ago [7,8] established the link, and recent accurate work by Palmer and Munro [9] has confirmed the exact relationship by direct testing and shown how the noise corresponds exactly to the T.E.

Since most companies flatly refuse to believe that there is a direct link between noise and T.E., it is common for companies to re-invent the wheel by testing T.E. and cross-checking against testbed noise checks. This is apparently very wasteful but has the great advantage of establishing what T.E. levels are permissible on production, as well as giving people faith that the test is relevant. For this learning stage of the process it is simplest to borrow or hire a set of equipment to establish relevance before tackling a capital requisition or to take sets of gears for test to the nearest set of equipment. Unfortunately, those few firms who have T.E. equipment usually use it very heavily so it may be better to ask a university if equipment can be hired. Newcastle [10], Huddersfield [11], and Cambridge [12] in the U.K.,

Ohio State University [13] and other researchers [14, 15, 16] have developed their own T.E. equipment and are usually happy to provide experience as well as a full range of equipment and analysis techniques. Academic equipment based on off-line analysis is often, however, not suited to high speeds or mass production.

References

1. Lemanski, A. J., Gear Design, S.A.E., Warrendale 1990. Ch 3.
2. Buckingham, Earle, Analytical mechanics of spur gears, Dover, New York. 1988.
3. Harris, S.L., 'Dynamic loads on the teeth of spur gears.' Proc. Inst. Mech. Eng., Vol 172, 1958, pp 87-112.
4. Gregory, R.W., Harris, S.L. and Munro, R.G., 'Dynamic behaviour of spur gears.' Proc. Inst. Mech. Eng., Vol 178, 1963-64, Part I, pp 207-226.
5. Munro, R.G., 'The Effect of Geometrical Errors on the Transmission of Motion Between Gears.' I. Mech. E. Conf. Gearing in 1970, Sept. 1970, p 79.
6. Cremer, L., Heckl, M., and Ungar, E.E., Structure-borne sound. Springer-Verlag, 1973, Berlin.
7. Kohler, H.K., Pratt, A., Thompson, A.M. Dynamics and noise of parallel axis gearing. Inst. Mech. Eng. Conf. Gearing in 1970, Sept, pp 111-121.
8. Furley, A.J.D., Jeffries, J.A. and Smith, J.D., 'Drive Trains in Printing Machines', Inst. Mech. Eng. Conference, Vibrations in Rotating Machinery, Cambridge, 1980, pp.239-245.
9. Palmer, D. and Munro, R.G., 'Measurements of transmission error, vibration and noise in spur gears.' British Gear Association Congress, 1995, Suite 45, IMEX Park, Shobnall Rd., Burton on Trent.
10. The Design Unit, Stephenson Building, Claremont Rd, Newcastle upon Tyne NE1 7RU, U.K. D.A. Hofmann.
11. Dept. of Mechanical Eng., Queensgate, Huddersfield, HD1 3DH, U.K. Prof R.G. Munro.
12. University Eng. Dept., Trumpington St., Cambridge CB2 1PZ, U.K. Dr J.Derek Smith.
13. Ohio State Univ., Mech. Eng. Dept., 206 West 18th Ave., Columbus, Ohio, 43210-1107. Prof D. R. Houser.
14 INSA de Lyon, Villeurbane, Cedex, France. Mr D. Remond.
15. University of New South Wales, Australia. Mr R.B. Randall.
16. Tech. Univ. of Ostrava, CZ - 703 88 Ostrava, Czech Republic. Mr. Jiri Tuma.

2

Harris Mapping for Spur Gears

2.1 Elastic deflections of gears

The basic geometric theory for spur gears assumes the "unwrapping string" generation of a perfect involute. We can then replace the two mating involute curves with a string unwrapping from one base circle and coiling onto the other base circle as in Fig. 2.1.

A contact between one pair of mating teeth should then travel along the "string," the "pressure line" or "line of contact" until it reaches the tip of the driving gear tooth. To achieve a smooth take-over, before one contact reaches the tip there must be another contact coming into action, one tooth space behind. For the theoretical ideal of a rigid gear the only requirement for a smooth take-over is that the base pitch, the distance between two successive teeth along the pressure line, should be exactly the same for both gears.

Unfortunately, although gear teeth are short and stubby, they have elasticity and there are significant deflections. The deflection between two teeth is partly due to Hertzian contact deflections, which are non-linear, but mainly due to bulk tooth movement because the tooth acts as a rather short cantilever with a very complex stress distribution and some rotation occurs at the tooth root. A generally accepted Figure for the mesh stiffness of normal teeth is 1.4×10^{10} N/m/m or 2×10^6 lbf/in/in, a Figure used by Gregory, Harris and Munro [1] in the late 1950s but one which has stood the test of time. As a rough rule of thumb we can load gears to 100N per mm of face width per mm module so a 4 mm module gear 25 mm wide might be loaded to 10,000N (1 ton). This load infers a deflection of the order of $400/1.4 \times 10^7$ m or 28.6 μm (1.1 mil).

Experimental measurement of this rather high stiffness has proved extremely difficult both statically and dynamically even with spur gears so that we are mainly dependent on finite element stressing software packages to give an answer. There is a significant effect at the ends of gears since the ability to expand axially reduces the effective Young's modulus and high angle helical gears have reduced contact support at one end and additional buttressing at the other end.

13

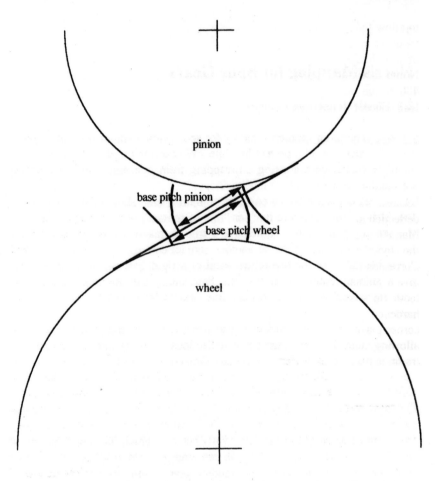

Chapter 2

Fig 2.1 Handover of contact betweeen successive teeth.

     Different manufacturing methods produce different root shapes and affect stiffness, but the main variations arise from variation of pressure angle or undercutting and, to a lesser extent, from low tooth numbers.

     The stiffness of each tooth varies considerably from root to tip, but with two teeth the effects mainly cancel. The highest combined stiffness occurs with contact at the pitch points and the stiffness decreases about 30% toward the limits of travel but the decrease is highly dependent on the contact ratio and gear details.

     In practice it is unusual for the applied load to be completely even across the face width as this implies that helix and alignment accuracies, and gear body deflections, must sum to less than a few μm. As a result, we have to allow for typically up to 100% overload and deflection at either end of the tooth, or in the middle if crowned, so deflections can be large. Using the rule of thumb that conventional surface-hardened teeth may be loaded to 100 N/mm facewidth/mm module, the above 4mm module gear (6 DP) loaded to 400 N/mm would deflect 400/14, i.e., 28 μm, nominally but, allowing for load concentrations, this could rise to 50 μm (2 mil).

2.2 Reasons for tip relief

Since there is deflection of the mating pair of teeth under load, it is not possible to have the next tip enter contact in the pure involute position because there would be sudden interference corresponding to the elastic deflection and the corner of the tooth tip would gouge into the mating surface. Manufacturing errors can add to this effect so that it is necessary to relieve the tooth tip (Fig. 2.2) to ensure that the corner does not dig in. Correspondingly, at the end of the contact, the (other) tooth tip is relieved to give a gradual removal of force. High loads on the unsupported corner of a tooth tip would give high stresses and rapid failure, especially with case-hardened gears which might spall (crack their case). In addition a sharp corner plays havoc with the oil film locally as the oil squeezes out too easily allowing metal to metal contact and accelerated failure. Tip relief design was traditionally a black art but can be determined logically.

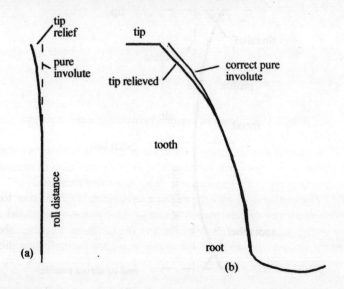

Fig 2.2 Picture of tip relief showing deviation from an involute in (a) and typical tooth shape (b).

The amount of "tip relief" needed in the example above can be estimated by adding the worst case elastic deflection, for example, 28.6µm + 70% (to allow for misalignment), to the possible base (adjacent) pitch errors of 3 µm on each gear and to the possible profile errors of 3 µm on each gear. The total tip relief needed is then 61 µm (2.5 mil). There can be some extra tip relief correction required if there is a large temperature differential between two mating gears, as one base pitch grows more than the other due to thermal expansion, but the effect is usually very small [2].

This "tip relief" can be achieved by removing metal from the tip or the root of the teeth or from both. There are two main schools of thought. The traditional approach was to give tip and root relief, as indicated in Fig.. 2.3, with a rather arbitrary division between the two and with the tip and root relief meeting roughly at the pitch point. The actual shape of the relief, as a function of roll angle, which is directly proportional to roll distance, tends to be almost parabolic.

There are two problems with this approach. It is not immediately clear where the tip of the mating tooth will meet the lower part of the working flank so it is more difficult to work out how much the effective root relief is at the point where the mating tip meets the flank. Rather more important is the fact that this parabolic shape of relief is not desirable from either noise aspects and for helical gears is undesirable from stressing aspects.

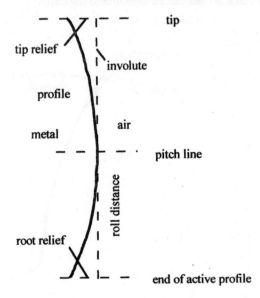

Fig 2.3 Tip and root relief applied on a gear.

In practice, we usually wish to have relief varying linearly with roll angle, starting at a point on the flank well above the pitch point so that there is a significant part of a tooth pair meshing cycle where two "correct" involutes are meeting.

When discussing profile corrections there are initially two uncertainties about the specifications. The first is whether the relief quoted is in the tangential direction or whether in the direction of the line of action. As the difference is normally only 6% on standard gears it is not important but most traditional profile measuring machines measure normal to the involute (i.e., in the direction of the line of action) and it is the movement or error in this direction that gives the vibration excitation so we usually specify this. When using a 3-D coordinate measuring machine it is again better to work in the direction of the line of action.

The other possible uncertainty is determining the position of a point up the tooth flank. The obvious choices of distance from root or tip are irrelevant as the profile ends are not accurate.

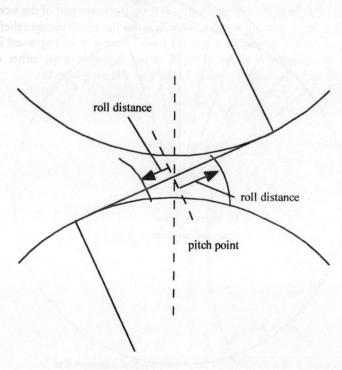

roll distance

roll distance

pitch point

Fig 2.4 Unwrapping string model.

Specifying actual radius is of little help in locating the correct points and referencing them to gear rotation. What is done in practice is to work in terms of roll distance. See Fig. 2.4. As the gear rotates and the "unwrapping string" leaves one gear base circle and transfers to the other there is a linear relationship between rotation and the distance that the common point of contact moves along the line of action. Roll distance is simply roll angle in radians times base circle radius. We measure and specify position in the tooth mesh cycle by giving the distance that the point of contact has travelled. Tooth flank starting and finishing points are unclear so design works in roll distance measured from the pitch point.

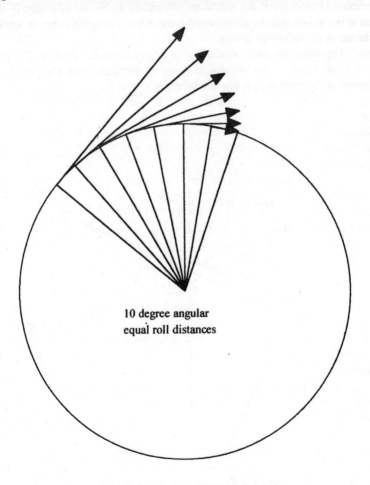

10 degree angular
equal roll distances

Fig 2.5 Effect of equal steps of roll on involute.

There is not a linear connection between roll distance and distance up the flank as can be seen from Fig. 2.5 which shows the "string" unwrapped at equal angular intervals and so equal distances along the line of action. Up the flank the distance intervals (between arrow tips) steadily increase.

When giving experimental measurements of profile or of the design on a single gear of a pair it is usual to show the reliefs relative to a perfect involute which is a straight vertical line up the page. Roll distance is vertical and the reliefs (to large scale) are shown horizontally as in Fig. 2.3. However when we are looking at the meshing of a pair of teeth the picture is turned on its side as in Fig. 2.6 so that roll distances are horizontal and reliefs are vertical. There can be problems locating exactly where on an experimental profile measurement the pitch point occurs as it can only be located by an accurate knowledge of the pitch radius and this depends on the centre distance at which the pair of gears will run.

The main choice in profile design is between giving both tip and root relief on the pinion so that the wheel (or annulus) stays pure involute for easy production or giving tip relief, but no root relief, on both, which is easier to assess and control. This choice can be controlled by production constraints of availability of suitable gear machines and cutters. In this book it is assumed that tip relief is given on both gears but there is no root relief to complicate the geometry.

A very special case arises for very large slow gears which have been in service for a while so that both pinion and wheel have worn away from their original (involute) profile. The most economical repair is then to leave the wheel as it is and adjust the profile of the pinion to suit the now incorrect wheel.

2.3 Unloaded T.E. for spur gears

Fig. 2.6 (a) shows diagrammatically what happens when we take two mating spur gear teeth, each with tip relief extending a third of the way down (but no root relief), and mesh them. All distances along the profile are in terms of roll distance, not actual distance, and so are proportional to gear rotation (multiplied by base circle radius).

The horizontal line represents the pure involute and the two tooth profiles, shown slightly apart for clarity, follow the involute profile to above their pitch line where they are relieved. In this case the tip reliefs are linear, as is modern custom. The combination of two teeth with perfect involutes in the centre is to give zero T.E. for this part of the mesh. Where there is tip relief it is irrelevant which gear has it as either gives a drop in the T.E. trace for the combination.

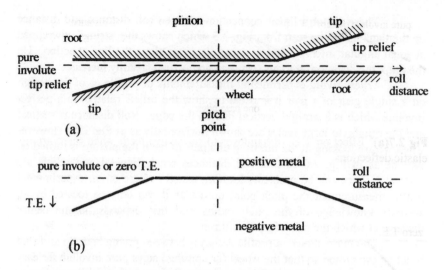

Fig 2.6 Effects of mating two spur gear profiles, each with tip relief.

T.E. traces are conventionally drawn with positive metal giving an upward movement but when testing experimentally the results can correspond to positive metal either way so it is advisable to check polarity. In the metrology lab this can simplest be done by passing a piece of paper or hair though the mesh.

The combined effect of one pair of teeth meshing under no load would be to give a T.E. of the shape shown in Fig. 2.6(b) with about one third of the total span following the involute for both profiles and generating no error. The tip reliefs then give a drop (negative metal) at both ends. The same effect is obtained if the relief is solely on the pinion at tip and root. However, the geometry is more complex at the root as the mating tip does not penetrate to the bottom of the machined flank.

Putting several pairs of teeth in mesh in succession gives the effect shown in Fig. 2.7(a). If there are no pitch or profile errors and no load applied so no elastic deflections, the central involute sections will be at the same level (of "zero" T.E.) and part way down the tip relief there will be a handover to the next contacting pair of teeth. One base pitch is then the distance from handover to handover. When we measure T.E. under no-load conditions we cannot see the parts shown dashed since handover to the next pair of teeth has occurred.

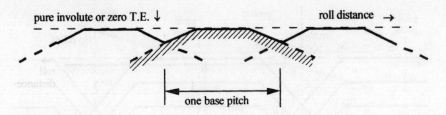

Fig 2.7(a) Effect on T.E. of handover to successive teeth when there are no elastic deflections.

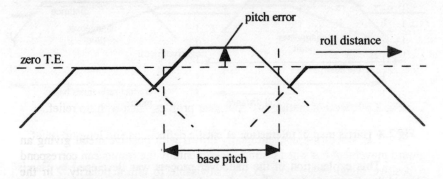

Fig 2.7(b) Effect of pitch error on position of handover and T.E.

Fig. 2.7(b) shows the effect of a pitch error which will not only give a raised section but will alter the position at which the handover from one pair to the next occurs,

2.4 Effect of load on T.E.

We wish to predict the T.E. under load as this is the excitation which will determine the vibration levels in operation.

As soon as load is applied there are two regimes, one around the pitch point where only one pair of teeth are in contact and one near the handover points where there are two pairs in contact, sharing the load but not, in general, equally. The total load remains constant so, as we are taking the simplifying assumption that stiffness is constant, the combined deflection of the two pairs in contact must equal the deflection when just one pair is in contact. In particular, exactly at the changeover points, the loads and deflections are equal if there are no pitch errors so each contact deflection should be half the "single pair" value.

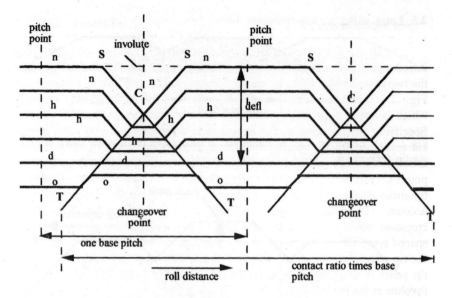

Fig 2.8 Harris map of interaction of elastic deflections and long tip relief.

This explanation of the handover process was developed by Harris [3] and the diagrams of the effects of varying load are termed "Harris maps." Fig. 2.8 shows the effect. The top curve (n) is the T.E. under no load and then as load is applied the double contact regime steadily expands around the changeover point. Curve (h) is the curve for half "design" load. At a particular "design load" the effects of tip relief are exactly cancelled by the elastic deflections (curve d) so there is no T.E. There is a downward deflection (defl) away from the "rigid pure involute" position but, as the sum of tip relief and deflection is constant, it does not cause vibration.

Above the "design" load the single contact deflections are greater than the combined double contact plus tip relief deflections. The result is as shown by curve (o) with a "positive metal" effect at changeover. Varying stiffness throughout the mesh alters the effects slightly, but the principle remains. In this approach it should be emphasised that "design" load is the load at which minimum T.E. is required, not the maximum applied load which may be much greater.

Since the eventual objective is to achieve minimum T.E. when the drive is running under load, there will normally be a desired design T.E. under (test) no-load. This leads to the curious phraseology of the "error in the transmission error," meaning the change from the desired no-load T.E. which has been estimated to give zero-loaded T.E.

2.5 Long, short, or intermediate relief

In 1970, Neimann in Germany [4] and Munro in the U.K. introduced and developed the ideas of "long" and "short" relief designs for the two extreme load cases where the "design" load is full load or is zero load. Fig. 2.8 shows the variation of T.E. with load for a "long relief design" which is aimed at producing minimum noise in the "design load" condition. Specifying the tip relief profile begins with determining the tip relief at the extreme tip points T, making the normal assumptions about overload due to misalignment and manufacturing errors. The necessary relief at the crossover points C (where contact hands over to the next pair of teeth at no-load) is half the mean elastic deflection and here we do not take manufacturing errors into account. Typically the relief at T may be 3 to 4 times that at C. The crossover points C are spaced one base pitch apart and the tip points T are spaced apart the contact ratio times a base pitch. It is, of course, simplest if the tip reliefs (which should be equal) are symmetrical. The start of (linear) tip relief is then found by extending TC backwards till it meets the pure involute at the point S.

An alternative requirement is to have a design which is quiet at no load or a very light load since this is likely to occur for the final drive motorway cruising condition or when industrial machinery is running light, as often happens.

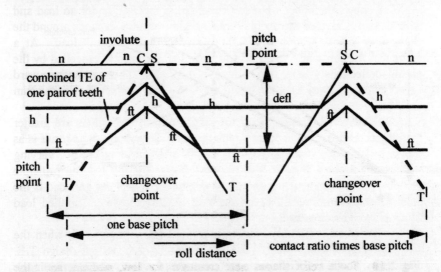

Fig 2.9 Harris map of deflections with a "short" tip relief design.

The "design" condition is zero load so we require "short relief" as shown in Fig. 2.9, which shows the variation of T.E. with load for "short" tip relief.

The pure involute extends for the whole of a base pitch so there is no tip relief encountered at all at light load (n). The tip relief at T must, however, still allow for all deflections and errors.

As load is applied we are then exceeding "design" load of zero and there will be considerable T.E. with high sections at the changeover points. Curve "ft" is the full torque curve where there is a section at changeover with double contact and hence half the deflection (defl) from the pure involute that occurs near the pitch points. Palmer and Munro [5] succeeded in getting very good agreement between predicted and measured T.E. under varying load in a test rig to confirm these predictions.

Care must be taken when discussing "design load" in gearing to define exactly what is meant because one designer may be thinking purely in terms of strength so his "design" load will be the maximum that the drive can take. If, however, noise is the critical factor, "design load" may refer to the condition where noise has to be a minimum and may be only 10% of the permitted maximum load. If the requirement is for minimum noise at, for instance, half load, then the relief should correspondingly be a "medium" relief. The short or long descriptions refer to the starting position of the relief, but the amount of relief at the tip of each tooth remains constant.

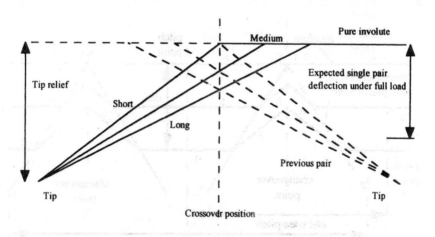

Fig 2.10 Tooth relief shapes near crossover for low, medium, and high values of design quiet load in relation to maximum load.

Fig. 2.10 shows for comparison the three shapes of relief near the crossover point for the conditions of the design quiet condition being zero, half and full load. For standard gears with a contact ratio well below 2 it is only possible to optimise for one "design" condition but as soon as the contact ratio exceeds 2 then there can be two conditions in which zero T.E. is theoretically attainable.

References

1. Gregory, R.W., Harris, S.L. and Munro, R.G., 'Dynamic behaviour of spur gears.' Proc. Inst. Mech. Eng., Vol 178, 1963-64, Part I, pp 207-226.
2. Maag Gear Handbook (English version) Maag, CH8023, Zurich, Switzerland.
3. Harris, S.L., 'Dynamic loads on the teeth of spur gears.' Proc. Inst. Mech. Eng., Vol 172, 1958, pp 87-112.
4. Niemann, G. and Baethge, J., 'Transmission error, tooth stiffness, and noise of parallel axis gears.' VDI-Z, Vol 2, 1970, No 4 and No 8.
5. Palmer, D. and Munro, R.G., 'Measurements of transmission error, vibration and noise in spur gears.' British Gear Association Congress, 1995, Suite 45, IMEX Park, Shobnall Rd., Burton on Trent.

3

Theoretical Helical Effects

3.1 Elastic averaging of T.E.

A spur gear, especially if an old design, will give a T.E. with a strong regular excitation at once per tooth and harmonics (Fig.. 3.1), even when loaded. The idea of using a helical gear is that if we think of a helical gear as a pack of narrow spur gears, we average out the errors associated with each "slice" via the elasticity of the mesh by "staggering" the slices.

If we have a helical gear which is exactly one axial pitch wide, the theoretical length of the line of contact remains constant. Fig. 3.2(a) shows a true view of the pressure plane which is the 3-D "unwrapping band" that unreels from one base cylinder and reels onto the other base cylinder.

With a spur gear the contact "point" in end view, i.e., 2-D, appears as a straight line parallel to the axis, but with a helical gear in 3-D, the contact line is angled at the base helix angle α_b. As each section along the face width will be at a different point in its once-per-tooth meshing cycle, there will be an elastic averaging of errors giving reduced T.E. Fig. 3.2(b) shows that if the slices are staggered, the total amount of interference and force remains roughly constant. In practice, using a helical gear is found to improve matters but not as much as might be hoped.

The idea is right but the realities complicate life since we can rarely get the axial alignment of two helical gears accurate enough. There are four tolerances involved even before we start thinking about elastic effects on gear bodies, supporting shafts, bearings and casing.

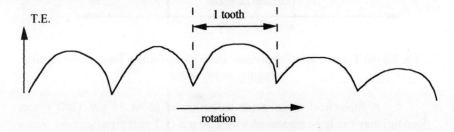

Fig 3.1 Typical section of T.E. of meshing spur gears.

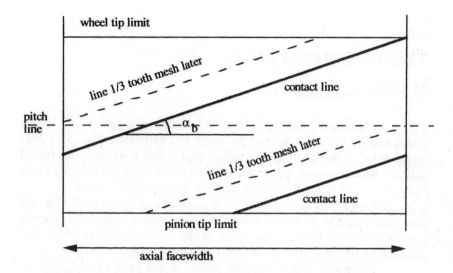

Fig 3.2 (a) View of pressure plane of helical gear showing contact lines.

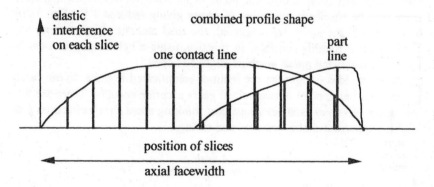

Fig 3.2 (b) Total of interferences on slices along contact lines summing to a roughly steady value.

A theoretical mean mesh deflection of about 15 μm (200 N/mm loading) may easily be associated with a 30 μm (1.2 mil) misalignment over a 150 mm (6 inch) face width. Hence an angular error of 2 in 10,000 still gives 100% overload at one end and zero loading at the other. With this variation in load the elastic averaging effects along the helix are much less effective and the helical gear transmission errors start to rise toward those of a spur gear.

Increasing helix angle so that there are several axial pitches in a face width improves the elastic averaging effect under load but penalties exist in increased axial loads and lower transverse contact ratios.

3.2 Loading along contact line

Another major effect with helical gears is indicated in Fig. 3.3 which is a view of a single tooth flank showing a contact line across the face. As the mesh progresses, the contact line comes onto the tooth face at the lower right corner, extends and travels across the face, and then disappears off the top left corner. With this engagement pattern there is no longer the necessity to achieve a smooth run-in with tip relief because we can do it with end relief. In a high power gear such as a turbine reduction gear a typical tooth face is much wider (axially) than it is high. This can give us a large strength bonus as the full loading per unit length of line of contact can be maintained nearly up to the tips of the teeth.

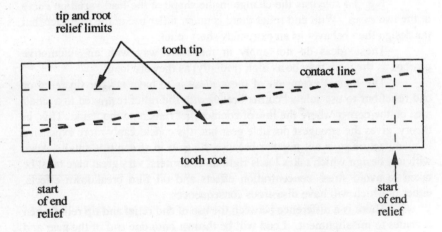

Fig 3.3 Theoretical flank contact line on a helical tooth face.

There is less tooth face "wasted" as a result of tapering in over two-thirds of a module at each end of the tooth, compared with more than a module (in roll distance) at top and bottom if the gear is designed as a spur gear. A chamfer is needed at the tooth tips as it is also needed at the end faces of a spur gear to prevent corner loading which gives very high local stresses and gives oil film failure. This stress relief chamfer is small in extent compared with (long) tip relief which can come one third of the way down the working flank.

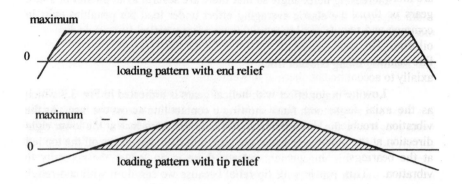

Fig 3.4 Variation of loading intensity along contact line with end and tip reliefs.

Fig. 3.4 suggests the change in the shape of the load variation curve in the two cases. With end relief there is much fuller use of the gear face, but the design then behaves as an extremely short relief.

These ideas do not apply in the same way with an automotive gearbox as the teeth may be as high (radially) as they are long.

The logical extension of these ideas is to have neither tip relief or end relief but to use solely "corner" relief with the relief restricted to a small area on the corner where the line of contact first runs onto the flank. This, in theory, gives the strongest possible gear but it is considerably more expensive to manufacture so is not popular because the gain in strength is small. Also, with any design which takes loads right up to corners, very great care must be taken to avoid stress concentration effects and oil film breakdown effects, either of which will have disastrous consequences.

There is a difference between the use of end relief and tip relief when it comes to misalignment. Load will be thrown onto one end of the gear and the effect will be similar to having a spur gear, so if a tip relief design has been used it is more likely to be quiet at higher loads. If end relief has been used, the profile will be much nearer a pure involute (an extremely "short" relief) and is likely to give relatively low T.E. at light loads but, correspondingly, a higher T.E. at design load.

3.3 Axial forces

Single helical gears produce axial forces which for a given torque are proportional to the tangent of the base helix angle of the gears. Axial forces are usually coped with easily in small gearboxes, but in large gearboxes there are more likely to be bearing limitations so it is common to use double helical gears or thrust cones to take out the axial forces. Thrust cones are not common and require skill to get the details right so that there is a satisfactory oil film. The local rigidity of the thrust flange must be carefully controlled or line contact will occur and one gear, usually the pinion must be able to move axially to accommodate thermal movements.

Low helix angles of less than 10° give relatively low axial forces so as the axial forces are about 1/6th of the radial we would expect little vibration trouble. Unfortunately, most gearcases are rigid in the radial direction at the bearings but often are relatively flexible in the axial direction at the bearings. This means that small forces may give disproportionate vibration. This problem is relatively easily identified when the drive is running by mode shape measurements and can often be solved simply by thickening or ribbing the bearing support plates. A 10° base helix angle with a 4 mm normal module gear requires a minimum face width of $4\pi \cos 20$ cosec 10° or 68 mm for good design so narrow gears will be pushed to higher helix angles. In general, the most difficult helical gears to design are those with narrow facewidths well below any possible axial pitch.

The higher axial forces that result with increase of helix angle will increase axial bearing loads and axial vibration excitation for a given T.E. In contrast, the higher helix angles will generally reduce T.E. so it is extremely difficult, if not impossible, to predict whether or not a change will give improvement. Spur gears, of course, produce no axial excitation but usually have a much higher T.E. unless a high contact ratio (greater than 2) design is used.

3.4 Position variation

One possible cause of vibration occurs when the force between the gears is constant and acts in a constant direction but oscillates sideways. This is a major cause of noise with Wildhaber-Novikov or Circ-Arc gears as the force application point moves a large distance.

Involute spur gears should not suffer from this problem if well aligned, but helicals may to a lesser extent. If we take a nominal contact ratio of r and look at the theoretical contact line lengths, we get the two extreme positions shown in Fig. 3.5.

These show the pressure plane for the worst case with a (correct) face width of an axial pitch and a small helix angle. This simple analysis ignores any end relief effects or tip relief effects and assumes a constant loading along the contact line. Practical teeth tend to give slightly larger effects.

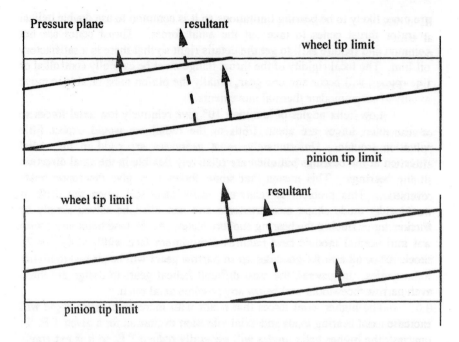

Fig. 3.5 Extreme positions of contact lines in pressure plane showing how the forces at the centres of each section of the contact line give a resultant force whose position varies.

The extreme position of the centre of action of the resultant force is determined by taking moments about one end and is approximately $(r-1)^2/2r + 1/2r$ which is $[(r-1)^2+1]/2r$ from one end. This has a minimum when r is $\sqrt{2}$ and the centre of force oscillates about .086 of the face width on either side of the centre of the face.

There is a corresponding radial force variation at the bearing housings of the order of 8% of the mean value when the gears are well supported close in or less if the supporting shafts are long. Although this effect exists in theory it is small and is dominated by axial force effects and conventional T.E. effects. Tip relief has a further complicating effect since all loadings near root and tip are reduced and with an older design there is a relatively concentrated load which runs along the pitch line.

Methods to reduce this effect have been proposed (by Rouverol [1]) but the reduction in effective flank area is significant and increases nominal stresses and the "silhouetting" takes little account of the complexities of real tooth profiles with tip and end reliefs. Increasing axial overlap reduces the effect but it is probably not worth considering for most gear noise problems.

If, in practice, experimental measurements suggested that bearing excitations at either end of a pinion were 180° out of phase, then the possibility of position variation excitation should be checked.

3.5 "Friction reversal" and "contact shock" effects

In the case of spur gears there was, at one time, a considerable body of academic opinion that ascribed much of the vibration of gears meshing to "pitch line friction reversal excitation." The theory said that there was effectively "dry" friction between gear teeth and that the direction of relative sliding between the gear teeth suddenly reversed at the pitch point and reversed when one pair of teeth left contact and the next pair started. This would give rise to a force in the sliding direction with an amplitude of ± the friction force and a roughly square waveform and much of spur gear noise was attributed to this effect. In addition there were assumed to be "sudden" shocks associated with gear teeth coming into contact and taking up load.

If this simple friction picture applied then with spur gears under an average load equivalent to 20 μm elastic deflection, a friction coefficient of 0.05 would give an oscillating force which was equivalent to ± 1 μm excitation in the sliding direction. Typically, however, the T.E. might be ± 2 μm in the pressure line direction and dominates the theoretical friction effects. The reality is considerably more complicated and the effects are much smaller because:

  (a) The effect of tip relief is to give a very gradual increase in the force between the teeth extending nearly to the point where the "friction" reverses. This relatively "gradual" increase in force (along the pressure line) gives a corresponding gradual increase in friction force unless specially desiged gears are used to exaggerate the effect.

  (b) The friction is not "dry" but elastohydrodynamic and so there is a slow viscous transition through the pitch point as the velocity reverses. This prevents the generation of sudden shocks from "friction reversal" and experimental investigations during work on Smiths shocks could not detect any such effects of sudden force changes [2].

Detailed experiments carried out by Houser, Vaishya and Sorensen [3] using accurate gears varied contact pressure, surface finish, lubrication and speed to investigate excitation in the direction normal to the line of pressure. The results showed motions in this direction comparable in size to the vibrations in the pressure line direction. Deduction of the excitations which were involved was however difficult due partly to the inevitable cross interactions which occur with any bearing system and partly due to the differences in effective response stiffnesses in the pressure line and normal directions as these can differ by a factor of 100. As might be expected the tribological conditions which are most likely to give either very thin oil films or limited metal to metal contact are the conditions which give high friction and associated vibration.   These conditions should be avoided as far as possible in service as they are also the conditions associated with surface failure mechanisms such as micropitting.

As far as "contact shocks" are concerned, when gears come into contact there is a rather small closing velocity between the mating flanks in the normal direction that is two orders smaller than the sliding or rolling velocities at the contact. With reasonable assumptions about the tip relief shape the estimated stress wave levels are small so there are negligible engagement shocks.   Typically 50 μm of tip relief will be taken up in about one third of a tooth interval so at a tooth frequency of 500 Hz the closing velocity v is about 75 mm/s. The corresponding stress wave intensity [4] is of the order of E v/c at the source and is $210 \times 10^9 \times 0.075/5000$ in steel or 3 MPa. On a contact area of 20 mm^2 this is only 20 N compared with a typical force variation due to T.E. of the order of $3 \times 10^{-6} \times 10^9$ or 3 kN so it is negligible.

These theoretical predictions have been borne out by practical measurements on extremely quiet gears by Munro [5] as well as by direct shock measurements on gears during work on condition monitoring [2] which showed no shocks at either entry or pitch points when the gears were operating correctly without asperity contact.   Shocks as small as 2 N occurring for only 20 microseconds could easily be detected by the test system used.

When we come to helical gears there are the same arguments that the friction forces change smoothly rather than abruptly. In addition, there is the major effect that roughly half the contact is occurring on either side of the pitch line so the corresponding friction forces are in opposite directions and tend to cancel out. The combination of effects means that "friction reversal" excitation may be ignored completely for helical gears and is small for spur gears. Similarly, "contact shock" effects are negligible for spur gears, and for helical gears, which have a very gradual take-up of force, the effects are small unless there is serious misalignment.   High contact ratio spur gears (see chapter 13) have the sliding contact friction forces opposing and cancelling each other at all points in the meshing cycle and so in theory can only generate net friction forces if there are serious accuracy errors.

3.6 No load condition

It is generally stated without thought that helical gears will always be quieter than spur gears but this is a dangerous assumption. It is certainly true that if there is good alignment between the gear helices in position and there is high loading then the elastic effects will even out errors and the mesh will be quiet.

In use however gears are liable to be loaded to much lower torques than their maximum load especially in automotive drives and in industrial machinery may spend much of their working day idling. Design loading is typically 100 N / mm / mm facewidth so for a 2 mm module gear the design load would be 200 N / mm and the corresponding elastic deflection about 15 μm (0.6 mil). At a typical working condition of one third load the theoretical mean deflection is only 5 μm so, as it is very easy to get misalignments much higher than this, the loading will be predominantly at one end of the teeth.

Contact for only perhaps a half or a third of the facewidth means that the theoretical vibration advantages of elastic averaging with helical gears will not be achieved as the pair will behave more as spur gears though with a design profile that has assumed full contact along the helix. Problems will also occur with heavily loaded gears that have been designed with high helix corrections to get even loadings at full torque but there will be little deflection or windup at light load so all the contact will be concentrated at the outboard end of the teeth and there will be little helical averaging effect.

References

1. Rouverol, W.S. and Pearce, W.J. 'The reduction of gear pair traansmision error by minimising mesh stiffness variation.' AGMA Paper 88-FTM-11. New Orleans, October 1988.

2. Smith, J.D., 'A New Diagnostic Technique for Asperity Contact.' Tribology International, 1993. Vol 26, No 1, p 25.

3. Houser, D.R., Vaishya, M, and Sorensen, J.D. 'Vibro-acoustic effects of friction in gears: an experimental investigation.' A.S.M.E., paper 2001-01-1516, 2001.

4. Roark's Formulas. 6th edition. Young, W.C., McGraw-Hill, New York, 1989, section 15.3.

5. Munro, R.G. and Yildirim, N., 'Some measurements of static and dynamic transmission errors of spur gears.' International Gearing Conf., Univ of Newcastle upon Tyne, September 1994.

4

Prediction of Static Transmission Error

4.1 Possibilities and problems

If we already have a gearbox available and can run it slowly under design torque, without wrecking gears or bearings, then the reliable and the most straightforward approach is to measure the quasi-static T.E. (see Chapter 7). This gives a very reliable answer with an accuracy of a fraction of a micron and can give major clues if anything is going badly wrong.

However, at the design stage it is desirable to have an idea of what the reaction of the design will be to the inevitable manufacturing errors, as far as noise and stress are concerned. Conversely, if an existing box is tested, it is an advantage to know what errors might have produced a given (undesirable) result. There is a fundamental problem as mentioned in section 1.5 that a dozen effects, each of possibly 2 μm metrology uncertainty, combine to give an answer, the T.E., which should be better than 1 μm uncertainty.

It is worth noting that in practice the most critical accuracy is the profile, so it is worth taking extra care with this measurement. When the profile is being measured with a conventional 3-D co-ordinate measuring machine, we must allow for all the errors on two axes (x and y) so it is difficult to achieve better than about 3 μm accuracy despite the manufacturer's claims. If we take the trouble to position a particular flank as shown in Fig. 4.1 we can improve accuracy by at least a factor of 2. The co-ordinates of the pitch point on the flank are simply $(r_b, r_b\tan\phi)$ and the gear does not have to be exactly in position. The gain in accuracy arises because there is very little movement in the y direction as the profile is traversed, so errors in this axis are minimal and movements in the x direction have very little effect on the (small) y corrections so errors in x are unimportant. Accuracy can then be better than 1 μm.

Despite the practical uncertainties of manufacturing accuracies, misalignments, and deflections, it is very worthwhile to use a simple computer model to check whether a design is or is not tolerant of errors, to assess the relative importance of the various errors on noise and stresses and to set realistic limits.

37

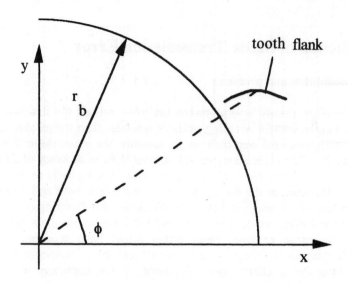

Fig 4.1 Improving profile accuracy by tooth position.

A model to check this does not necessarily have to be extremely accurate, perhaps within a few percent, but it should be able to give a quick, cheap comparative assessment of different designs with realistic assumptions.

There are full 3-D finite element programs which take a gear tooth and calculate the deflections when it meshes with another gear tooth but such models are extremely complex. Since there can easily be of the order of 10,000 node points in a realistic model with perhaps 1000 boundary points defining a tooth shape, the calculations are large and, equally important, there is a large technical effort required each time to enter each set of boundary conditions. This level of effort is justified for very high performance (expensive) gearing [1], especially if "corner" relief is used, but is uneconomic for normal industrial gearboxes. A simpler, cheaper model of gear tooth meshing is needed compatible with practical realities of permissible computer and software costs and the practicability of the amount of input information required.

4.2 Thin slice assumptions

A suitable model to choose in practice as a compromise is the "thin slice" model. The helical gear is assumed to act as if it were a pack of thin 2-dimensional slices with each slice of tooth behaving independently of its neighbours and deflecting solely due to the contact forces on that slice.

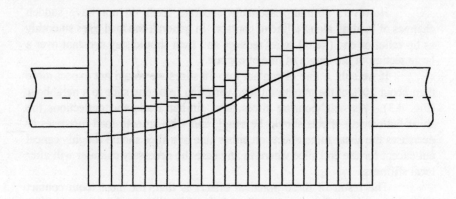

Fig 4.2 Sketch of thin slice model of helical gear.

Exactly whether we are better to take the "slices" in the transverse plane and assume that each slice is restrained axially by its neighbours, or whether we think of a slice local to a tooth and normal to the tooth, is a problem which is open to argument.   Fortunately it makes negligible difference for helical gears with low helix angles so it is simpler to think in the transverse plane, as sketched in Fig. 4.2.   The principal theoretical objection to the thin slice model is "buttressing" because we know that if we apply a load at one point only on a tooth the local deflections are less than the "thin slice" estimate because the neighbouring slices support or "buttress" the slice, due to shear stresses and to longitudinal bending stresses.

At the end of a tooth the local stiffness is significantly lower because there is no support from the outboard end, as well as the effective modulus being lower due to axial expansion reducing Poisson's ratio effects.

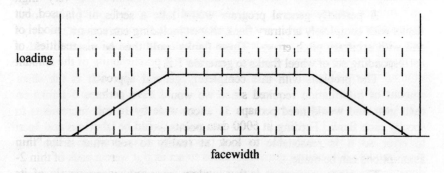

Fig 4.3 Typical variation of load between gear 'slices.'

However, normal reasonably accurate teeth do not have sudden changes of loading along a line of contact. In general, the load rises smoothly as tip relief or end (helix) relief reduces and then should stay constant over a large section of the length of line of contact.

If we split a line of contact into 30 slices we would not expect more than about 20% of the maximum load variation from one slice to a neighbour (Fig. 4.3). As neighbouring slices have similar loads and deflections the shear buttressing effects should be small, and with smooth load increases or decreases the shear force effects on either side of a slice should roughly cancel out except for the end slice where in any case the necessary chamfer will alter local stiffness.

The result of these practical effects is that, for most tooth contact lines, buttressing effects are small and the thin slice model is much more accurate than might be expected. One time that buttressing effects are significant is when one gear is much wider than the other and no end relief has been given. This condition, of course, tends to cause rapid failures at the sharp corner because of stress concentration effects and because lubrication is impossible at a sharp corner. Differing gear widths tends to occur with small pinions which have been cut directly into a shaft to give minimum diameter. Another area where buttressing is important occurs with high helix angle gears which are too narrow to have end relief, where one end of the tooth flank is less supported due to the angle of the end of the tooth. Even in this case, the extra stiffness of one tooth end may largely compensate for the lower stiffness of the mating tooth end to give roughly constant mesh stiffness. However, the local root stresses will be much higher with the unsupported tooth end.

4.3 Tooth shape assumptions

A perfectly general program would take a series of pinion tooth flanks with completely arbitrary flank shapes including corrections and errors and with arbitrary pitch errors. These flanks could then be matched with a corresponding set of wheel flanks to generate T.E.

The problem with this completely general approach is the sheer amount of information required since we would have perhaps 6 flanks on each gear and would need perhaps 31 slices wide by 16 roll increments to specify each flank. Feeding in 6000 data points would be laborious and open to error so it is reasonable to look at reality to see what simplifying assumptions can be made.

The main assumption is that modern, reasonably accurate machines will be used for production. Such machines, whether hobbers, grinders or shavers have the characteristic that they produce a surprisingly consistent profile shape on the tooth flanks. Shapers produce a less consistent flank shape but are also relatively less used. The flank shape which is produced is consistent within about 2 μm (< 0.1 mil) and, as our standard measurement techniques are only correct to about 2 μm at best, we are justified in assuming that all profiles on one side of the teeth are effectively the same "as manufactured." They will probably not be the correct profile, due to machine or cutter or design errors, but they will be consistent. In position in the drive however, the apparent errors may vary due to eccentric mounting or swash.

The second corollary to using a modern hobber or grinder is that true adjacent pitch errors will be small, typically less than 3μm at worst. As measured they may appear to be greater if there is a large eccentricity. If we take a "perfect" 20 tooth gear and mount it with an eccentricity of ± 25 μm (1 mil) a pitch checker will record an adjacent pitch "error" ranging up to 7.8 μm as shown in Fig. 4.4 (a).

The maximum apparent error obtained is *eccentricity* x 2 sin (180/N) where N is the number of teeth. This "error" is fortunately not a real error which will affect the meshing due to the beneficial properties of the involute.

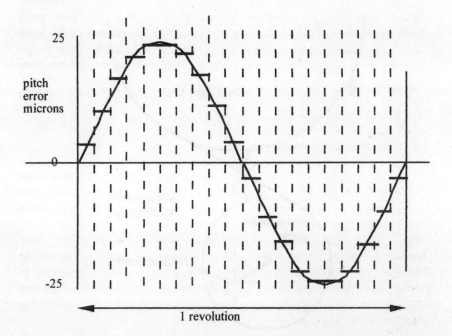

Fig 4.4(a) Spurious readings of adjacent pitch error due to eccentricity.

The all-important base pitch has not been altered by the eccentric mounting of the gear so the required smooth handover to the next tooth pair will not be affected. This apparent adjacent pitch error due to eccentricity is a problem which causes great concern and produces a large number of spurious "theoretical" deductions about once per tooth (and harmonics) noise effects. In practice, as indicated in Fig. 4.4 (b), mating a "perfect" wheel with a "perfect" but eccentric pinion will give a smooth sinusoidal T.E., not the staircase effect of large once per tooth errors with step changes at changeover. This is because the fundamental conjugate involute "unwrapping string" theory still applies even though the centre of the base circle is moving relative to the wheel centre.

The other important factor in relation to adjacent pitch errors is that they cannot give significant vibration generation at once-per-tooth frequency and harmonics. This, at first sight, seems peculiar and if, as in Fig. 4.5, we plot typical random adjacent pitch errors around a pinion, it is not obvious why once-per-tooth frequency cannot exist.

The mathematics of a series of random height (pitch) steps *of equal length* gives the result that there is no once-per-tooth or harmonics (see Welbourn [2]).

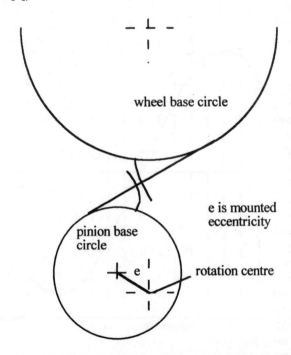

Fig 4.4(b) Effect of eccentric pinion mounting on transmission smoothness.

adjacent pitch error

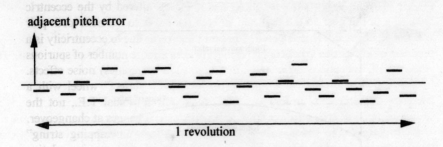

1 revolution

Fig 4.5 Typical adjacent pitch error readings.

The restriction of equal length steps is valid for modern gearing and only breaks down with extremely inaccurate gears of old design. The result can be seen more straightforwardly if we integrate the N adjacent pitch errors since the integral of adjacent pitch is cumulative pitch which sums to zero round a full revolution of N teeth. As the integral of N values is zero, the integral of all the fundamental components must be zero. And so there are no components at N, $2N$, $3N$, etc. times per revolution. The mathematics ties in with the experimental observation that pitch errors do not give the steady whines associated with once per tooth excitations, but do give the low frequency graunching, grumbling noises that we associate with relatively inaccurate gearboxes with high pitch errors.

Again, as adjacent pitch errors in good manufacturing are small and their contribution to steady noise at any given frequency is even smaller (< 0.5 μm at worst), we can afford to ignore their effect on noise. This assumption is curiously pessimistic since pitch errors can have the positive effect of breaking up steady once-per-tooth whines. On some drive systems, such as inverted tooth chains, it is a standard trick to introduce deliberate random pitch errors to produce a more acceptable noise. The effectiveness of this approach is partly due to a slight real reduction in sound power level at tooth frequency, and partly due to the complex non-linear response of human hearing.

The standard methods of manufacture tend to give a profile which is consistent along the axial length of the teeth but the helix matching between two mounted gears is rarely "correct" along the tooth. In some cases there may be helix correction to allow for the pinion body bending and twisting under the imposed loads. More commonly, there is no attempt to correct exactly for distortion but there are end reliefs, crowning, and misalignment so an analysis needs to allow for these. There may also be helix distortions associated with long gears expanding thermally more in the middle than at the ends, which are better cooled.

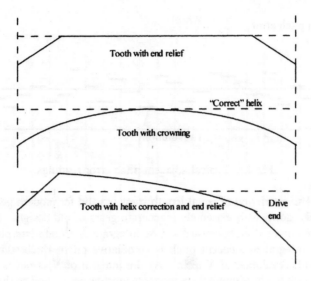

Fig 4.6 Different helix corrections.

In this discussion, end relief is used to describe a relief which is typically linear and is restricted to a short distance at either end of the helix, whereas crowning applies over the whole face width and is parabolic (or circular) with the relief proportional to the square of the distance from the gear centre (see Fig. 4.6).

Specifying the (consistent) profile is predominantly a question of specifying the tip reliefs on wheel and pinion. Old designs tended to give a tip relief extending down to the pitch line and roughly parabolic, so the relief was roughly proportional to the square of the distance from the pitch line. This form of tip relief is very easily computed but as it gives rather noisy and highly stressed gears, it is little used in modern designs. The more common linear relief starts abruptly from a point which is typically a roll distance about one third of a base pitch from the pitch point. There is negligible root relief if both wheel and pinion tips are corrected, but root relief also must be used if only one gear is corrected.

4.4 Method of approach

Fig. 4.7 shows a schematic view of the pressure plane for a pair of helical gears. The x direction is the axial direction and y is along the pressure plane in the direction of motion of the contact points.

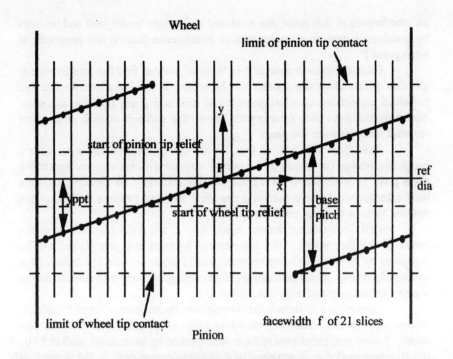

Fig 4.7 View of pressure plane.

The reference diameter is more commonly called the pitch line and is where the two pitch cylinders touch. The pressure plane is limited at either end as the "unwrapping band" unreels from one base cylinder and reels onto the other base cylinder. Within the pressure plane, contact can only occur in a limited strip since contact must cease when the teeth tips are reached, however high the load. In practice, however, the effect of tip relief is usually to taper off contact before the geometric tip limit is reached.

On any given tooth flank, contact can only occur on a single contact line which runs at an angle α_b (the base helix angle) to the axial direction. However, there may be contacts on previous or later tooth flanks which are still within the contact zone. Fig. 4.7 has been drawn for the case where the contact pattern is symmetrical and one contact line is running through the pitch point P at the centre of the face width and on the pitch line (where the two pitch cylinders touch). This central point P is the reference point $x = 0$, $y = 0$ from which all measurements of position in the pressure plane are made. If contact occurs anywhere along the pitch line $(y = 0)$ there is (by definition) no tip relief on either gear as all profile corrections are measured relative to the profile at the pitch point. There will, in general, be contact and an interference at this point due to elastic deflections under load and we start by arrbitrarily assuming an amount of interference (ccp in the program) at pitch point P.

Once the interference at P is "known" we can find the interference at all other points along the contact line by adding in the extra interference due to helical corrections or misalignment and subtracting any tip relief amounts. Summing the local slice interference times slice stiffness at each point gives the total force between the gears.

This force will not, at first, be the correct desired force but with a rough knowledge (or guess) of the overall contact stiffness we can correct the pitch point interference to get a better answer and carry on iterating until the total interference force is within a specified amount, perhaps 0.05%, of the applied force in the base pitch direction.

Helix corrections depend solely on x, the axial distance from the centre of the face width. The interference between the gear flanks will be increased by bx where b is the relative (small) angle between the helices, due to manufacturing misalignments together with gear body movements due to support deflections and body distortions.

Crowning will reduce the interference by an amount crrel * $(x/0.5f)^2$ where crrel is the amount of crowning relief at the ends and f is the face width. Linear end relief also reduces interference by an amount endrel * (x - 0.5 ff), provided this is positive (or 0 if negative); endrel is the amount of end relief and ff is the length of face width that has no end relief. Fig. 4.8 shows the effects.

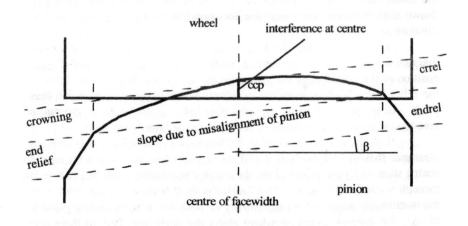

Fig 4.8 Sketch of effects of reliefs and misalignment on helix match.

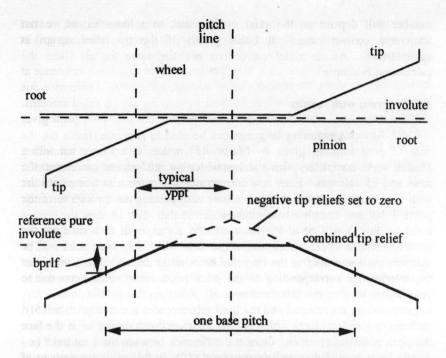

Fig 4.9 Modelling tip relief corrections on a single mesh.

Tip relief corrections for a slice depend upon the distance (yppt) of the contact point from the pitch line. Fig. 4.9 shows two teeth with tip relief, shown slightly spaced away from the horizontal line which represents the true involute (on both gears).

The resulting combined tip relief is shown in the lower part of the diagram and can be modelled easily by putting the tip relief to be bprlf * (|y| - position of start of relief)/(0.5 p_b - position of start of relief) where bprlf is the relief at the ±0.5 p_b handover position (at zero load) and p_b is the base pitch. All negative values of tip relief, those near the pitch point, are put to zero to correspond to the central "pure involute" section.

Two further factors need to be considered when estimating the extra clearance that will be given by tip relief. The first is that the contact on the centre slice will move away from the pitch point P as the mesh progresses through a complete tooth cycle so that if the base pitch is p_b and we divide the meshing cycle into 16 (time) steps, each step will add p_b/16 to all values of y the distance of the slice contact point from the pitch line and so influence the tip relief. The second is that in addition to the contact line which runs roughly through the pitch point P, there will be other contact lines 1 or 2 base pitches ahead and 1 or 2 base pitches behind. The exact number will depend on the axial overlap and, to a lesser extent, on the transverse contact ratio. It helps greatly if the tip relief design is symmetrical. As tip relief corrections are the same for all slices the calculation is simple.

4.5 Program with results

Any programming language can be used to generate results but the ease of programming given by Matlab [3] makes it a strong candidate. Matlab works completely with matrices which for this calculation consist of 5 rows and 25 columns. Each row corresponds to a particular line of contact with row 3 as the one which starts at time zero passing not through the pitch point P but one complete base pitch earlier so that after 16 steps the central point on line 3 will be at P. Each column corresponds to a slice and an arbitrary choice of 25 slices across the face width has been made. The matrices corresponding to the tip relief helix relief are added to a matrix of the interference corresponding to the pitch point interference between the gear bodies to give the interference at all points on the contact lines. Any negative values are rejected and the local interferences are multiplied by local stiffness to give total force which is then compared with design force to adjust the pitch point interference. Once the difference between the total force and design force drops below an arbitrary level (50N in this case) the pitch point interference is recorded, and the mesh is incremented one sixteenth of a base pitch for the next step of the 32 that correspond to two-tooth mesh cycles.

Transmission Error Estimation Program

```
% Program to estimate static transmission error
% first enter known constants or may be entered by input
facew=0.125;      % arbitrary 25 slices wide gives 5 mm per slice
baseload = input('Enter base radius tangential applied load     ');
bpitch=0.0177;    % specify tooth geometry 6mm mod
misalig=40e-6;    % total across face                line 4
bprlf=25e-6; % tip relief at 0.5 base pitch from pitch point
strelief = 0.2;      % start of linear relief as fraction of bp from pitch pt
tanbhelx=0.18;    % base helix angle of 10 degrees
tthst = 1.4e10; %  standard value of tooth stiffness
relst=strelief*bpitch; % start of relief                          line 9
ss = (1:25);hor = ones(1,25);     % 25 slices across facewidth
x = (facew/25)*(ss - 13*hor);  % dist from facewidth centre
crown = (x.*x)*8e-6/(facew*facew/4);  % 8 micron crown at ends
ccp = 10e-6 ;  % interference at pitch pt in m at start
```

```
%  alternatively ccp = baseload/ facew*tthst
te = zeros(1,32);                    %                              line 12
for k = 1:32 ;        % complete tooth mesh 16 hops        **************
   for adj = 1:15          % loop to adjust force value >>>>
for contline = 1:5 ;      % 5 lines of contact possible? $$$$$$$$$$$$$
yppt(contline,:)=x*tanbhelx+hor*(k-16)*bpitch/16+hor*(contline-3)*bpitch;
rlief(contline,:)=bprlf*(abs(yppt(contline,:))-relst*hor)/((0.5-strelief)*bpitch);
posrel = (rlief(contline,:)>zeros(1,25)) ;% finds pos values only
actrel(contline,:) = posrel.* rlief(contline,:);% +ve relief only
interf(contline,:)=ccp*hor+misalig*x/facew-actrel(contline,:)-crown;  % local
%   interference along contact line
posint = interf(contline,:)>0 ; %   check interference positive
totint(contline,:)=interf(contline,:).*posint ;            %    line 23
end            % end contact line loop        $$$$$$$$$$$$$
% disp(round((1e6*totint)'));pause % only if checking interference pattern
ffst = sum (sum(totint));   % total of interferences
ff = ffst * tthst * facew /25;  % tot contact force is ff
residf=ff- baseload ;         % excess force over target load
%   disp(residf) ; pause    % only if checking
   if abs(residf) > baseload*0.005;                    %   line 27
   ccp = ccp - residf/(tthst*facew) ;      % contact stiffness about 1e9
   else
   break            % force near enough
   end
  end             % end adj force adjust loop  >>>>>>>
   if adj==15;                                          % line 33
   disp('Steady force not reached')
   pause
   end
te(1,k) = ccp * 1e6;   % in microns
intmax(1,k) =max(max(totint));            % maximum local interference
end %   next value of k                **********************
xx = 1:32;                           % steps through meshline  40
peakint = max(intmax) ;           % max during cycle
contrati = 1.6 ;   % typical nominal contact ratio
stlddf = peakint*facew*contrati*tthst/baseload ;% peak to nominal
disp ('Static load distribution factor') ; disp(stlddf) ;
figure;plot(xx,te);xlabel('Steps of 1/16 of one tooth mesh');
ylabel('Transmission error in microns');
```

In the program the first 10 lines (not counting % comment lines) set up the constants and an arbitrary starting position of 10 μm interference at the central pitch point. Line 12 generates the crowning relief proportional to distance x (from face centre) squared. Line 14 starts the main loop to do the 32 steps corresponding to 2 complete tooth meshes. Line 15 starts the force adjustment loop which is set arbitrarily to 15 convergences. Normally the loop will converge to within 1% of the applied force (roughly 0.05 micron) well before 15 tries and will break out in line 31. If not, a warning is displayed and the program is stopped.

Instead of guessing an arbitrary starting interference ccp (10 μm) it should be a better guess to take baseload / facew times nominal contact ratio times tthst. The problem with this is that if there is high crowning or large misalignments or tip reliefs, we do not know what the effective length of line of contact is.

Along each line of contact (line 16) the distance (yppt) of each x slice contact from the pitch line is the sum of the base helix effect, the movement due to the 32 steps and the movement due to the change from one contact line to the next. The tip reliefs are calculated in line 18, and those that are positive detected in line 19 so that the negative ones can be put to zero in line 20.

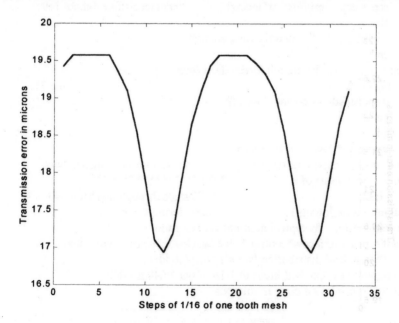

Fig 4.10(a) Predicted static T.E. result for 40 μm misalignment.

Line 21 sums the effects of body interference, misalignment, tip relief and crowning, then in line 23 only the positive interference values are retained. All values of interference are summed and multiplied by the slice stiffness to give the total contact force ff which would only be correct if the initial value of ccp was correct.

This force is compared with the desired contact force and the difference is divided by a guessed overall mesh stiffness to adjust the pitch point interference ccp to a new "better" value. The loop repeats until the agreement is within 50 N (11 lbf) in this case. Finally, the next step of the 32 steps is selected and convergence is fast because the starting value of ccp will be nearly correct.

A typical result from this program is shown in Fig. 4.10(a) for the design figures in the program and a contact load of 20,000N (2 tons). The average value of deflection is due to elastic tooth deflections and is ignored since it is only the vibrating variation that is important for noise purposes. The T.E is about 3 μm p-p.

Fig. 4.10(b) is similar but is for only 10 μm misalignment and though the peak to peak T.E. is similar the waveform is better so there will be smaller harmonics.

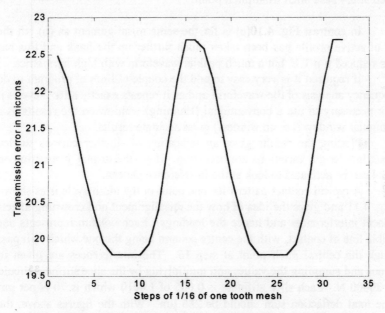

Fig 4.10(b) Predicted static T.E. result for 10 μm misalignment.

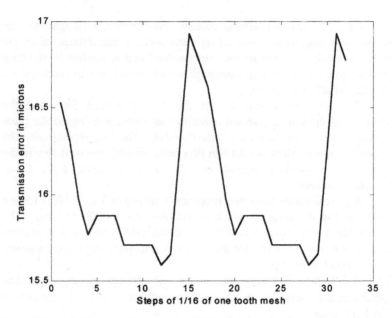

Fig 4.10(c) Predicted static T.E. result for 10 μm misalignment with relief started at 0.4 base pitch from pitch point.

In contrast Fig. 4.10(c) is for the same misalignment as (b) but the start of active profile has been taken much further up the flank and this has given reduced p-p T.E. but a much peakier waveform with high harmonics.

If required it is very easy to add the couple of lines of program to do a frequency analysis of the waveform and as it repeats exactly after 2 cycles it is not necessary to use a conventional (Hanning) window on the results as a rectangular window (i.e. no window) gives accurate results.

Plotting the results gives an indication of whether curious sudden contact line length variations are occurring. If so, the display instruction on line 24 can be activated to look at the interference pattern.

A typical contact pattern for one point in the mesh cycle is as shown in Fig. 4.11 and gives the idea of how the misalignment and crowning affects the local interferences and hence the loadings. Each column represents one possible line of contact, with the centre column being the one which will pass through the central pitch point at step 16. The interferences are given in microns and summing the values and multiplying by the slice stiffness should give 20000 N. Each slice stiffness is 0.005 of 1.4e10 which is 70 N per μm so the total deflection sum should be 286 μm. With the figures above, the rounding does not give the exact value.

0	0	0	0	0
0	0	0	0	0
0	0	0	0	0
0	0	0	2	0
0	0	0	5	0
0	0	0	6	0
0	0	0	8	0
0	0	0	9	0
0	0	0	11	0
0	0	0	13	0
0	0	0	14	0
0	0	0	16	0
0	0	0	13	0
0	0	0	11	0
0	0	0	8	0
0	0	0	5	0
0	0	0	3	0
0	0	1	0	0
0	0	7	0	0
0	0	12	0	0
0	0	18	0	0
0	0	24	0	0
0	0	30	0	0
0	0	35	0	0
0	0	37	0	0

Fig 4.11 Distribution of contact deflections.

4.6 Accuracy of estimates and assumptions

A simple program such as the one given will provide a very effective method of comparing different designs and, in particular, their sensitivity to misalignment and profile changes. The program not only gives the peak-to-peak of T.E. but also gives the maximum load per unit face width during the cycle which gives the static load distribution factor. This is the ratio of the actual peak loading to the nominal loading that would be obtained if the load spread evenly across the whole length of nominal contact line, roughly contact ratio times face width. This in AGMA 2001 is Cm (= Km) or Cmf * Cmt, or in DIN/ISO/BS is Khα * Khβ. The figures obtained for this ratio are often above 3, especially for relatively lightly loaded gears of old design, so that the gear is only taking a third of the load that could be taken at the same root and contact stresses, if the loading were evenly spread.

The factors that affect the accuracy of the estimates are:

(i) Profile and pitch manufacturing errors. These are surprisingly small, typically 2 μm which corresponds to 10% of a typical 20 μm tooth deflection. The effects on T.E. are much reduced due to elastic averaging across a helical gear but are more significant for stresses.

(ii) Alignment errors. These can be due to helix manufacturing errors but are much more likely to be due to the mounting errors of the gears on their spindles, the gearcase, or the bearings, especially with plain bearings. If there is poor design, such as overhung gears on slender spindles, then the gears can deflect very large amounts and alignment errors can easily exceed the tooth deflection. Gearcases which are not symmetrical can give different deflections at the bearings and so contribute to alignment errors. Crowning eases T.E. problems but at the cost of increasing stresses.

(iii) Tooth stiffness variation. Using the standard value of $1.4 * 10^{10}$ N/m/m for all conditions appears somewhat crude and an accurate figure requires many assumptions and a major finite element program, as well as a detailed knowledge of the tooth root shape. However, variation of tooth stiffness does not have a dramatic effect on T.E. or stresses. Teeth of standard form will vary relatively little in combined mesh stiffness because as one tooth flexes more towards the tip, the other is more rigid at its root. There is a variation as the contact nears the teeth tips and the stiffness reduces about 30%. In practice, we do not usually let the contact approach the tips with spur gears and the effect of tip relief is to start reducing the contact force well before the part of the mesh where the stiffness drops significantly. With modern helical gears the loadings may run further up the teeth but the helical effects average out the local stiffness variations so the T.E. is little affected. At the ends of the teeth there is a reduction in tooth stiffness but there should also be end relief (or crowning) reducing the force, and the effect is small (<10%) unless helix angles are very high.

(iv) 3-dimensional effects. A base helix angle of 10° gives axial forces less than 20% of the tangential forces. Take a wildly idealised gear mounting, as in Fig.. 4.12, with elastic deflections occurring due to the bearings and with gear diameter equal to the bearing span. The axial forces would give radial deflections at the bearings of the order of 0.1 F/k, where F is the radial force and k the bearing radial stiffness.

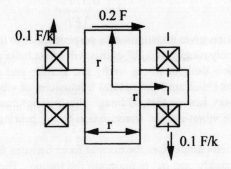

Fig 4.12 Effects of helical axial forces on alignments.

This would give axial deflection at the teeth of 0.1 F/k and a corresponding misalignment of about 0.1 * F/2rk, then a face width of r would give 0.05 F/k across the face width. The result is 10% of the lateral deflection of the gear attributable to the bearings which could be significant in those designs where support stiffnesses have been lowered to reduce internal natural frequencies.

Another effect can occur if supporting shafts are slender as the torque generated due to the axial component of contact force twists the gear as sketched in Fig. 4.13. With the dimensions shown of diameter equal to bearing span, the lateral forces at the shaft ends will be 0.1 F giving an angular rotation at the (narrow) gear of

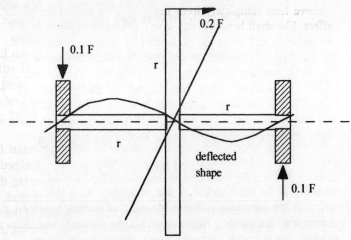

Fig 4.13 Tilting of gear due to helical axial forces.

$$\theta = \frac{0.1Fr^2}{3EI}$$

This rotation gives misalignments perpendicular to the direction of the helix so only roughly tan 20° of this will affect helix matching.

For the few designs where shafts are slender and allow high lateral deflections (>0.2 mm) the natural frequencies of vibration of the gears will be very low so that, although the misalignment will cause T.E. and hence vibration, the transmission to the bearing housings will be low.

As base helix angle rises the normal force between the gear teeth must rise by roughly sec α_b to maintain the torque. The length of contact line will, on average, also rise by sec α_b so the normal loading will remain roughly the same so the deflection of the teeth in the normal direction will remain roughly the same. The tangential (transverse) deflection will rise by sec α_b so the apparent stiffness in the transverse plane will then be reduced by a factor of cos α_b, but the effect is very small for normal helix angles.

(v) Gear body distortions. Gear bodies are not rigid so they twist and bend under loading, especially if either a pinion has a face width/diameter ratio approaching 1 (a "square" pinion) or if a wheel has a rim which is thin and distorts locally. For high performance gears, these distortions, together with any axis movements due to bending shafts, will be estimated and corrections applied to the helix to cancel out the expected deflections. Gears which have not been corrected will distort and, in extreme cases, may twist enough to remove load completely from the non-drive end of the pinion. This effect, like shaft bending, can give a dramatic increase in stresses and an increase in T.E. so the possibility of local distortion should always be checked. T.E. estimates which do not allow for major distortion or for shaft bending will be highly inaccurate. Corrections for pinion twist can become rather complicated in the general case but are simplified if we work backwards from an assumption that the tooth loading is constant. The twist angle is proportional to the square of the axial distance along the pinion and, at the free end is

$$T\,L/2GJ$$

where T is the total torque, L the facewidth, G is the shear modulus and J is the torsional stiffness moment of inertia, based on the root diameter of the pinion. Bending within the pinion is less likely to give trouble, but a rough estimate using simple beam theory is that the central deflection under an evenly distributed load W is

$$5W L^3/(384\ E\ I)$$

where L is the facewidth, E is Young's modulus and I is the bending moment of inertia which is probably best based on the pitch diameter. With highly loaded high facewidth pinions, which have been helix-corrected to even out stresses at maximum load, there is an inherent noise problem at low load because contact will be dominantly at the outboard end of the pinion and is liable to give high T.E.

It is sometimes possible with clever design to get pinion bending effects to partially cancel torsional windup effects.

(vi) Gear body movements. Corrections for shaft bending are usually small if the gear is supported symmetrically but can be substantial if a gear is overhung from bearings. They may be estimated roughly by adding the effects of bearing deflections, bending of the shaft outboard and bending of the shaft between the bearings using standard structures expressions. Fig. 4.14 shows the layout for an overhung gear. The final value for the slope of the gear is

$$\frac{W}{K}\frac{b}{a}\frac{1}{a}+\frac{W(a+b)}{Ka}\frac{1}{a}+\frac{Wb}{a}\frac{a^3}{3EI}\frac{1}{a}+\frac{Wb^2}{2EI}$$

where W is the load applied, K is the local bearing stiffness, a is the span between the bearings, b is the overhang to the centre of the gear, E is Young's modulus, and I is the local shaft bending moment of inertia.

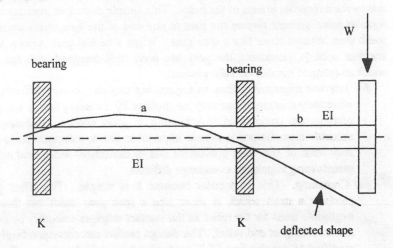

Fig 4.14 Overhung gear support shaft deflection.

These simple estimates are sufficient to see if the corrections are significant and whether it is necessary to bother with more detailed calculations and the accompanying manufacturing costs if helix correction is needed. A full analysis needs to take into account shear effects and pinion bending and can allow for variable loading along the contact line.

4.7 Design options for low noise

When designing standard spur gears for low T.E. there are few options since the only variable is the profile, assuming that the pitching is good as occurs usually. The possible approaches for normal contact ratios are

(i) If low load is of major importance, use "short" relief so that there is handover from pure involute to pure involute.
(ii) If high load is of major importance, use "long" relief as in section 2.5 with the tip relief at the changeover ± 0.5 p_b points equal to half the expected elastic deflection.

Both approaches can work reasonably well at their working load, provided design, manufacture, alignment, etc. are good. Howver they must give noise off-design and these spur gears will be sensitive to manufacturing errors.

The spur gear alternative is to use a nominal contact ratio above 2 to achieve a handover which is effectively "long" relief under full load and pure involute under light load. See Chapter 13.

Helical gears should be quieter than the corresponding spur gears due to the averaging effects of the helix. This simple deduction goes astray as soon as misalignment throws the load to one end of the face width since the mesh then behaves more like a spur gear. When a helical gear is noisy there are four options, (assuming the gear has been well designed with the face width an integral number of axial pitches):

(i) Improve alignment. Easy to suggest but this can be very difficult and ultimately alignment can only be checked by a blueing test or a copper plating test under load. Achieving a good enough alignment by accurate manufacture is almost impossible due to tolerance build-ups. Any form of gear axis movement due to deflection under load makes maintaining alignment even more difficult.
(ii) Crowning. This is popular because it is simple. The effect is to produce a mesh which is more like a spur gear mesh but there is negligible need for tip relief as the contact engages smoothly by using the crowning as end relief. The design profile can correspondingly be modified to get the best T.E. under load on a fairly narrow effective face width, like a spur gear. However, it is common to use crowning

with a profile which has no tip relief and gives very good T.E. at light loads with some penalty in T.E. at higher loads.

(iii) Heavy end relief. Like crowning, it is possible to use end relief together with a profile which is nearly pure involute. This acts like a spur gear giving low T.E. under light loads since there is an involute profile, but will give a reasonable T.E. at heavy loads since the length of contact line remains constant, providing that the effective face width is an integral number of axial pitches.

(iv) High contact ratio. As with spur gears, if the effective contact ratio is 2, inferring a nominal contact ratio of about 2.25, then the drive should be very quiet at low and high loads. See Ch 13.

Of these options (ii) has the disadvantage of giving high stresses whether or not the alignment is good whereas (iii) and (iv) only give high stress when there is severe misalignment. The ultimate design is probably to have a combination of (iii) and (iv) with a contact ratio only just over 2 and use blue checks to give reasonable alignment. In general, increasing helix angle gives a smoother drive but with the corresponding end thrust and axial vibration effects.

References

1. Houser, D.R., Gear noise sources and their prediction using mathematical models. Gear Design, SAE AE-15, Warrendale 1990, Ch 16.

2. Welbourn, D.B., 'Forcing frequencies due to gears.' Conf. on Vibration in Rotating Systems, I. Mech. E., London, Feb. 1972, p 25.

3. Matlab *The Math Works Inc.*, Cambridge Control, Jeffrys Building, Cowley Road, Cambridge CB4 4WS or 24 Prime Park Way, Natick, Massachusetts 01760.

5

Prediction of Dynamic Effects

5.1. Modelling of gears in 2-D

Static determination of T.E. under load is sufficient for most drives where the loading is relatively heavy and the inertias are low so that there is little danger of the length of line of contact varying greatly or of the teeth losing contact. The T.E. is then the input vibration and, as the system remains reasonably linear in its behaviour, it can be modelled using a conventional matrix approach in the frequency domain. Drives which are lightly loaded or which drive high inertias, such as printing rolls, may lose contact with rather dramatic results. It is then possible for the teeth to be in contact for less than 10% of the time with rather large impulsive forces while they are in contact. The simple assumption of a linear system with an input displacement of the quasi-static T.E. is then no longer realistic and a more detailed model is required (see section 5.2 and Chapter 11).

Even when the teeth do not come fully out of contact the simple assumption of a linear system can be wildly unrealistic. This is due to the large variations in the true length of the contact line, partly due to the gear flank shapes and partly due to the vibration. If the nominal mean elastic deflection in the mesh is of the order of 10 μm, then a vibration of 2 μm can easily alter the contact stiffness by a factor of 2 by changing the length of the line of contact during the vibration. A simple assumption that stiffness is proportional to nominal length of line of contact is near the truth for well-aligned spur gears but not true for misaligned gears, especially helicals.

The simplest realistic model of a pair of gears is shown in Fig. 5.1. Axial movements are negligible or ignored although the gears are taken to be helical. There is considerable simplification if we take the linear axis along the line of thrust and ignore any motion perpendicular as being small since it is only due to (small) friction effects which are in the main self-cancelling for helicals. Four degrees of freedom are involved, two linear and two torsional and if the system is linear with a constant contact stiffness s_c the estimation of response is simple.

A force P at the contact will give linear and torsional responses to each of the two gears. The relative movement d at P is the sum of the four responses together with the contact deflection due to the contact stiffness s_c and damping coefficient b_c.

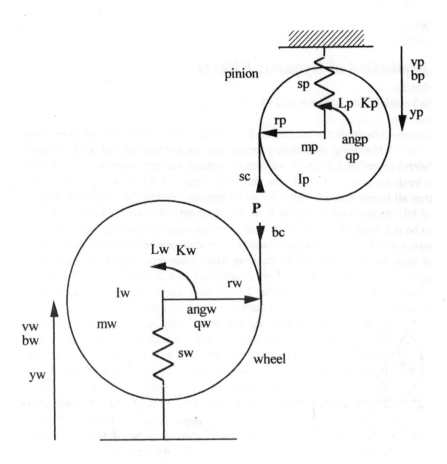

Fig 5.1 Simple 2-dimensional model of a gear pair vibration.

It is necessary to work from the common force to the deflections of the system since we cannot work from the combined deflection back to force.

$$d = P\left[\frac{1}{sp + j\omega bp - mp\omega^2} + \frac{rp^2}{kp + j\omega qp - Ip\omega^2} + \frac{1}{sw + j\omega bw - mw\omega^2} + \frac{rw^2}{kw + j\omega qw - Iw\omega^2} + \frac{1}{sc + j\omega bc}\right]$$

    This relative movement is the excitation, the T.E., so from d we can
determine P, the tooth force. Also if required we can determine the forces
transmitted through to the (rigid?) bearing housings.

    If it is necessary to determine the response for a two-stage gear drive
the problem becomes much more complicated. A two-stage box can be
sketched as shown in Fig.. 5.2 and as, in general, the lines of thrust for the
two meshes (A to B and C to D) will not be in the same direction we need to
use two co-ordinates for the position of the centre of each gear on the
intermediate shaft.

    The input and output gears can each be described with a single
lateral co-ordinate in the direction of the relevant line of thrust and of course
a torsional co-ordinate. It may be more useful to specify two co-ordinates so
that all lateral co-ordinates are x and y but this needs 12 co-ordinates instead
of 10. As there are 10/12 co-ordinates there are as many equations of motion
to be put down and a further two which determine the tooth forces P and Q in
terms of all the co-ordinates which contribute to the interference and the T.E.
at each mesh. A typical equation balancing external and D'Alembert forces
is:

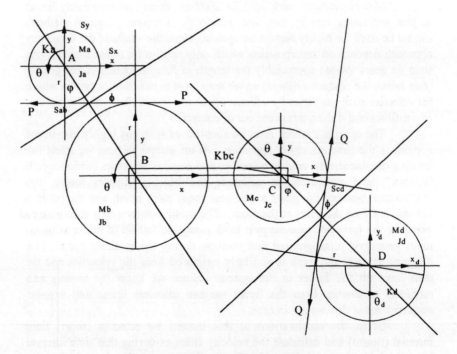

Fig 5.2 Model of two-stage gearbox.

$$y_b[S_{by} - M_b \, \omega^2] - P\sin(\varphi_{ab} - \phi_{ab}) - [M_c \, \omega^2 y_c + Q\sin(\varphi_{cd} + \phi_{cd})]\alpha_{bc}S_{by} = 0$$

In this equation S values are stiffnesses, P and Q are contact forces and α_{bc} is the response at C due to a unit force at B.

This is inevitably more complex than the analysis for a single stage, even without any complications from 3-dimensional (axial) effects which would increase the number of equations by roughly 50%. As the level of complexity rises considerably it is debatable whether the extra effort is worthwhile since there are uncertainties about many of the stiffness parameters. These stiffness uncertainties may be greater than the interaction effects between the stages and, as estimates of loss of contact are likely to be inaccurate due to lack of information about damping in impacts, we ignore two stage effects and concentrate on drives which can be isolated as a single stage and then idealised as in Fig. 5.1.

5.2 Time marching approach

Matrix methods work well for systems which stay reasonably linear so that stiffnesses vary by, say, less than 20%. Frequency domain methods cannot be used for highly non-linear systems since the whole of the frequency approach depends on superposition which only applies for linear systems. As soon as gears vibrate appreciably the length of line of contact varies greatly (and hence the contact stiffness) so we may have to deal with a system where the effective stiffness varies by a factor which may be 1000:1 within a fraction of a millisecond if the gears come out of contact.

The approach which must be adopted, as with any highly non-linear system, is the time marching approach. At an instant in time we select the existing displacements, angles, velocities and angular velocities (which are all "known") and use them to calculate the bearing support forces, the interference between the gears at the gear mesh pitch point, and the relative velocity between the gears at the mesh. The mesh interference is then used to calculate the force between the gear teeth using the full set of information on tooth geometry, misalignment and position during the meshing cycle. The damping force at the mesh is similarly estimated from the velocities and we then have all the forces in the system. Since we know the masses and moments of inertia, from the forces we can calculate linear and angular accelerations at this instant in time.

Given the accelerations at this instant we select a (short) time interval (timint) and calculate the velocity changes during that time interval by multiplying the accelerations by the time increment. We also calculate the corresponding displacement changes by multiplying the velocities by the time increment. This gives us the new velocities and displacements at the end of the time interval. These will be used for the force determinations for the next interval.

When computers were slow and lacking in memory this direct approach was too slow so it was necessary to indulge in complicated routines such as Runge-Kutta for interpolation and extrapolation to reduce computational effort. This is no longer necessary and it is simpler to take shorter time intervals to check accuracy or to ensure convergence.

5.3 Starting conditions

Any time-marching computation has to start from an arbitrary set of starting positions and velocities which will not be correct since they will not correspond to the steady vibration in the "settled-down" state. As we are starting from a "non-steady state vibration" condition there will be an initial starting transient which will take several cycles of vibration at each natural frequency to die away. The larger the initial error, the larger the transient will be and the longer will it take to die away to the point where one tooth meshing cycle is much the same as the next. We can guess roughly how long it will take for a vibration mode to die away by using the experimental observation that few modes have a dynamic amplification factor above 10. This infers a non-dimensional damping factor > 0.05 giving a decay of 25% per cycle so 10 cycles will reduce the transient to less than 5%.

It is not a good idea to set all starting values to zero since torsionally soft shafts will have to wind up (and deflect sideways) a large amount to take up the steady components of deflection to get bearing loads and shaft torques roughly right. This will take a long time before the system settles down.

We also have the fundamental problem of how to model a steady drive torque through the torsionally flexible input shaft, but if we simply put a pure torque on the end of a "light" shaft we remove the important effects of the torsional stiffness of the input shaft since the torque at the pinion remains constant. The alternative to using a steady input drive torque is to rotate the outboard end of the input shaft by an amount which will, on average, give the required input torque and keep this angular rotation (a pre-twist) fixed. The input torque will then vary slightly as the gears vibrate but the variation will be small. This modelling of the system is in good agreement with what happens in practice where there is often a very high referred moment of inertia at input and output of a gear drive system so high frequency torsional movements at the outboard ends of the input and output shafts are negligible.

The associated problem is that most drive systems are not tied to "earth" and are not prevented from rotating steadily. In mathematical terms they are "free-free" systems with a lowest natural frequency of zero. If we attempt to calculate the system as it is we are liable to find that, as in reality, it rotates steadily. This, although not disastrous, is inconvenient when we wish to look at results so we normally tie one part of the system to "earth", usually via a very flexible shaft so that the system displacements cannot wander off to infinity.

To find the "pre-twist" position of the input is reasonably straightforward since we can sum up the steady state angular movements due to the two shaft torsions, the two gear lateral deflections and the mesh deflection. In general, the mesh deflection is so small compared with shaft windups that it can be ignored. If we then start the sequence from the "static" position there will be initial transients but they will be small compared with the transients from a zero load position.

There is a complication in deciding when the system has "settled down" to a steady state because a non-linear vibrating system generally does not reach a state of steady vibration if contact is lost, but vibration amplitudes vary irregularly. Both the amplitude of bounce and the time between impacts varies so it is not as easy to decide when the starting transients have disappeared. Displaying, for example, a dozen tooth mesh cycles will usually show whether starting transients have decayed.

5.4 Dynamic program

```
% Matlab program to estimate forces under loss of contact. SI units.
clear;              % Enter known constants Damping must not be excessive
sp = 2e7; sw = 6e7; mp = 30; mw = 70;     % linear stiffn and masses
Kpr=4e6;Kwr=1.5e7;Iprr = 20; Iwrr=90;     % ang eff. stiffn and masses
bp = 1e3; bw = 2e3 ; qpr =1.5e2 ; qwr = 3e3 ;        % eff. damping coeffts.
tr= input('Enter pinion input torque divided by pinion base radius ');
freq = input('Enter tooth meshing frequency in Hz ');        % line 6
kk = round(20000/freq);               % steps for 1 tooth mesh
timint = 5e-5 ;               % time for single step 1/20000 sec
predefl = tr * (1/Kpr + 1/sp +1/sw + 1/Kwr);        % elastic defl.of shafts
% and torsions under steady torque referred to contact. then zero of
% input torsion is predefl from zero force position (ignores contact defl)
yp=-tr/sp;yw=-tr/sw;rthw=-tr/Kwr;rthp=-yp-yw-rthw;        % set initial
values
vp = 0 ; vw = 0 ; revp = 0 ; revw = 0 ;        % velocities at mesh        line 11
facew=0.105;bpitch=0.0177;        % specify tooth geometry 6mm mod +++
misalig=40e-6;bprlf=25e-6;        % relief at 0.5 base pitch from pitch point
strelief = 0.2;               % start linear relief as fraction of bp from pitch pt
slicew=facew/21;tanbhelx=0.18;tthst = 1.4e10 ;        % standard value
```

```
relst=strelief*bpitch;tthdamp = 1e5;    % eff.value at 10000 rad/s Q = 14+++
ss = (1:21);hor = ones(1,21);           % 21 slices  across face width      line 17
x = ss - 11*hor;                        % dist from face width centre in slices
for tthno = 1:20;                       % number of complete meshes
for k = 1:kk ;      % complete tooth mesh 20000/freq hops **************
ccp =  yp + yw + rthp + rthw ;              % interference at pitch pt in m
ccpv = vp + vw + revp + revw ;      % relative velocity between gears   line 22
for contl = 1:4 ;           % 4 lines of contact possible  $$$$$$$$$$$$$
yppt(contl,:)=x*slicew*tanbhelx+hor*k*bpitch/kk+hor*(contl-3)*bpitch;
rlief(contl,:)=bprlf*(abs(yppt(contl,:))-relst*hor)/((0.5-strelief)*bpitch);
posrel = (rlief(contl,:)>zeros(1,21)) ;
actrel(contl,:) = posrel.* rlief(contl,:) ;              % +ve relief only
interf(contl,:) = ccp*hor + misalig*x/21 - actrel(contl,:);       % local int
posint = interf(contl,:)>0 ;                  % check in local contact
equivint(contl,:) = interf(contl,:).*posint + posint*tthdamp*ccpv/tthst; % l 30
end                         % end contact line loop         $$$$$$$$$$$$$
ffst = sum (sum(equivint));         % force due to stiffness and damping
ff = ffst * tthst * slicew ;                    % tot contact force is ff
datp =k + (tthno - 1)*kk; fff(datp) = ff ;
if datp == 30; intmicr = round(equivint*1e6); disp(intmicr);
end                         % check on pattern              line 36
% total contact force    >>>>>>>>>>>>>>>>>>>>>>>>>>>> dynamics
accyp = -(ff + sp*yp + vp*bp)/mp;                % pinion acc.linear
accyw = -(ff + sw*yw + vw*bw)/mw ;               % wheel acc.linear
accthp = -(ff + (rthp-predefl)*Kpr + revp*qpr)/Iprr ; % pinion ang at mesh
accthw = -(ff + rthw*Kwr + revw*qwr)/Iwrr;       % wheel ang at mesh line 40
vp = vp + accyp * timint ; vw = vw + accyw * timint;        % velocities
yp = yp + vp * timint ;  yw = yw + vw * timint ;            % displ.
pdispl(datp) = yp*1e6;              % for monitoring pinion support force
revp = revp + accthp * timint ; revw = revw + accthw * timint;   % line 44
rthp = rthp + revp * timint ; rthw = rthw + revw * timint;    % ang displ
xt(datp) = datp /20;
end                         % next value of k ***************
end                                  % tthno loop end        line 48
figure;plot(xt,fff);xlabel('Time in milliseconds');
ylabel('Contact force in Newtons');
figure;plot(xt,pdispl);xlabel('Time in milliseconds');
ylabel('Pinion displacement in microns');
end
```

The program starts by setting up the gear body constants and asking for the mean contact load and the tooth meshing frequency. The original torsional stiffnesses are converted into equivalent linear stiffnesses K/r^2 at base circle radius and moments of inertia are turned into equivalent inertias I/r^2 again acting along the pressure line. Correspondingly, angles are multiplied by the relevant base circle radius to turn them into equivalent linear displacements rthp and rthw along the pressure line.

Lines 12 to 16 (not counting comment lines) specify the gear meshing parameters and figures for the tooth stiffness and the effective viscous damping between the teeth per unit length (while in contact), based on the Q (the dynamic amplification factor at resonance) being about 14 for vibration at 1600 Hz.

Line 19 then starts the sequence of, in this case, 20 tooth meshing cycles with each tooth mesh splitting into kk hops to make each roll distance step correspond to interval "timint." The calculation then proceeds in a manner similar to section 4.5, finding the all-important interference ccp at the pitch point and hence the interference pattern between the teeth on 4 lines of contact. The interference pattern (where positive) gives the elastic forces but also tells us where the teeth are in contact. Forces proportional to velocity are generated to add damping only where the teeth are in contact. In the program, this force is in the form of an extra effective interference proportional to damping coefficient times velocity divided by tooth stiffness (line 30).

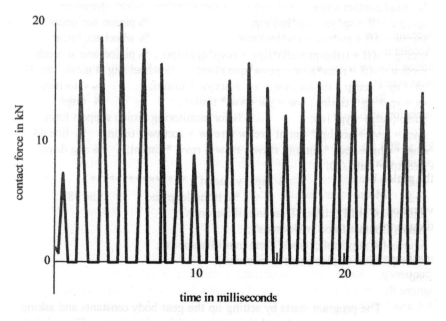

Fig 5.3 Prediction of contact force variation with time with helical gear.

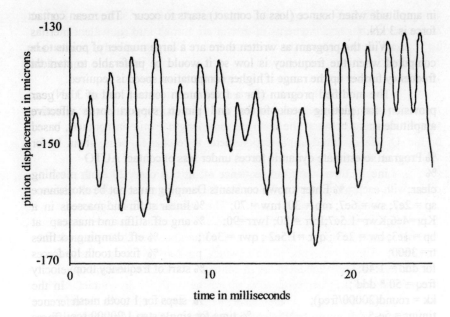

Fig 5.4 Prediction of variation of pinion displacement with time.

The total mesh force ff is generated in line 33 and is stored for plotting and to be used to calculate accelerations in lines 37 to 40. Accelerations and velocities are multiplied by the time increment and are added to existing values to give the new velocities and displacements for the next step of time.

Results from the program are shown in Fig. 5.3 for the contact force variation with time. The corresponding pinion vibration is shown in Fig. 5.4.

These are for an extreme case where the gears are lightly loaded (3 kN at 800 Hz tooth frequency) and are coming well out of contact. Once the pinion vibration is known, multiplying by the pinion support stiffness gives the pinion bearing vibrating forces.

Mean values are not important as it is only the variation that gives vibration and involute gears can tolerate considerable lateral deflections though they are highly sensitive to misalignments.

An extra loop can be put around the program to vary the tooth meshing frequency and extract the vibration or peak impact force for each frequency. Since initial conditions produce transients, it is necessary to ignore the first few milliseconds of response before extracting maxima. Figs. 5.5 and 5.6 show the results of such a program with the typical sudden jumps in amplitude when bounce (loss of contact) starts to occur. The mean contact force is 3 kN.

   With the program as written there are a large number of points to be computed when the frequency is low so it would be preferable to start the frequency further up the range if higher computation speed is required.

   The modified program (for a fixed mean contact load of 3 kN) and provision for plotting peak forces and pinion support force vibrating amplitude is:

```
% Program to estimate dynamic forces under loss of contact AUTO
%
clear;              % Enter known constants Damping must not be excessive
sp = 2e7; sw = 6e7; mp = 30; mw = 70;      % linear stiffn and masses
Kpr=4e6;Kwr=1.5e7;Iprr = 20; Iwrr=90;      % ang eff. stiffn and masses
bp = 1e3; bw = 2e3 ; qpr =1.5e2 ; qwr = 3e3 ;        % eff. damping coeffts.
tr= 3000;                                  % fixed tooth load
for ddd = 1:40;                            % start of frequency loop
freq = 50 * ddd ;
kk = round(20000/freq);                    % steps for 1 tooth mesh
timint = 5e-5 ;                 % time for single step 1/20000 sec
predefl = tr * (1/Kpr + 1/sp +1/sw + 1/Kwr);       % elastic defl.of shafts
% and torsions under steady torque referred to contact. then zero of
% input torsion is predefl from zero force position (ignores contact defl)
yp=-tr/sp;yw=-tr/sw;rthw=-tr/Kwr;rthp=-yp-yw-rthw;     % set initial values
vp = 0 ; vw = 0 ; revp = 0 ; revw = 0 ;       % velocities at mesh
facew=0.105;bpitch=0.0177;      % specify tooth geometry 6mm mod ++++
misalig=40e-6;bprlf=25e-6;      % relief at 0.5 base pitch from pitch point
strelief = 0.2;            % start linear relief as fraction of bp from pitch pt
slicew=facew/21;tanbhelx=0.18;tthst = 1.4e10 ;    % standard value
relst=strelief*bpitch;tthdamp = 1e5;      % eff.value at 10000 rad/s Q = 14++
 ss = (1:21);hor = ones(1,21);                 % 21 slices across facewidth
 x = ss - 11*hor;               % dist from facewidth centre in slices
for  tthno = 1:20;                          % number of complete meshes
for k = 1:kk ;           % complete tooth mesh 20000/freq hops ****
ccp =  yp + yw + rthp + rthw ;              % interference at pitch pt in m
ccpv = vp + vw + revp + revw ;              % relative velocity between gears
for contl = 1:4 ;               % 4 lines of contact possible  $$$$$$$$
yppt(contl,:)=x*slicew*tanbhelx+hor*k*bpitch/kk+hor*(contl-3)*bpitch;
rlief(contl,:)=bprlf*(abs(yppt(contl,:))-relst*hor)/((0.5-strelief)*bpitch);
posrel = (rlief(contl,:)>zeros(1,21)) ;
actrel(contl,:) = posrel.* rlief(contl,:) ;            % +ve relief only
interf(contl,:) = ccp*hor + misalig*x/21 - actrel(contl,:);    % local int
```

```
posint = interf(contl,:)>0 ;                          %   check in local contact
equivint(contl,:) = interf(contl,:).*posint + posint*tthdamp*ccpv/tthst ;
end                                 % end contact line loop    $$$$$$$
ffst = sum (sum(equivint));         % force due to stiffness and damping
ff = ffst * tthst * slicew ;                          % tot contact force is ff
datp =k + (tthno - 1)*kk;  fff(datp) = ff ;           % logs force to file
%  total contact force    >>>>>>>>>>>>>>>>>>>>>>>>>> dynamics
accyp = -(ff + sp*yp + vp*bp)/mp;             % pinion acc.linear
accyw = -(ff + sw*yw + vw*bw)/mw ;            % wheel acc.linear
accthp = -(ff + (rthp-predefl)*Kpr + revp*qpr)/Iprr ;        % pinion ang
at mesh
accthw = -(ff + rthw*Kwr + revw*qwr)/Iwrr;    % wheel ang at mesh
vp = vp + accyp * timint ; vw = vw + accyw * timint;  % new velocities
yp = yp + vp * timint ;  yw = yw + vw * timint ;       % new displ.
pdispl(datp) = yp*1e6;               % to check loop progress
revp = revp + accthp * timint ; revw = revw + accthw * timint;
rthp = rthp + revp * timint ; rthw = rthw + revw * timint;   % ang displ
end           %  next value of k  New values of displ, angles etc.*****
end                                  % tthno loop end
xzx(ddd) = 50 * ddd;
totno=length(fff);
stff(ddd) = max(fff(100:totno))/1000;     % peak after settling for 5 millisec
annal = fft(pdispl(100:totno)); fftno = length(annal);
brgvb(ddd) = 20*4* max(abs(annal(2:fftno)))/fftno;           % p-p value
clear pdispl fff
end                                         %main frequency loop
figure;plot(xzx,stff);xlabel('Frequency of excitation');
ylabel('Maximum contact force in kN');title('3000N mean load');
figure;plot(xzx,brgvb);xlabel('Frequency of excitation');
ylabel('Vibrating force through pinion bearing p-p');title('3000N mean load');
end
```

5.5 Stability and step length

The requirement for a short time interval in the computing arises from the necessity to calculate for a time short enough so that a large spring force or damping force is not allowed to "act" for so long that it over-corrects for a deflection or velocity and reverses the direction. In practice this means selecting a time interval which is not greater than one-tenth of the periodic time of the highest natural frequency encountered in the system. This can be found either from a linear analysis or guessed from the tooth stiffness and the effective masses of the gears.

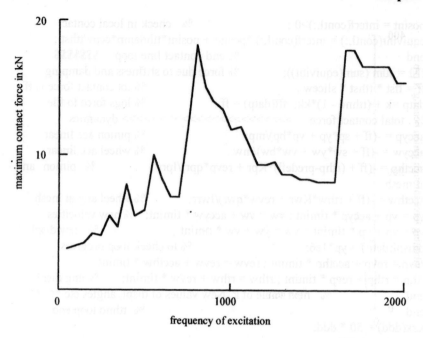

Fig 5.5 Prediction of variation of maximum contact force with tooth frequency.

In the example given, the highest natural frequency (when in full contact) is of the order of 1600Hz so with a periodic time of 600µs, a time interval of 50µs was taken. A test run with half the time interval (25µs) quickly checks that the computation is satisfactory since there is no significant change in the result.

The other factor which can give instability in a calculation is the use of a damping that is too high. Since we know that in a mechanical system damping is stabilising, there is a tendency to try a computation with a high level of damping on the assumption that the computation will then be stable. The opposite applies because the very high damping force acting for a finite time is liable to reverse the velocity giving instability. It is easy to apply too high a damping if the effect of the multiplication by ω is forgotten. The product of the damping coefficient and the contact natural frequency should be less than the mesh contact stiffness initially by a factor of about 10. As with high spring stiffnesses, reducing the time interval step helps to give stability. If problems are encountered the simplest approach is to reduce damping and time interval and if the system is still unstable to check the signs of all terms in the computation.

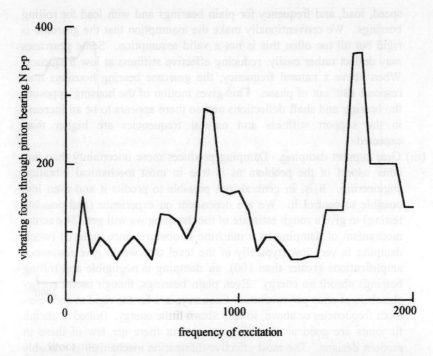

Fig 5.6 Variation of vibrating pinion support force with tooth frequency.

5.6 Accuracy of assumptions

Assessment of the accuracy of the assumptions made involves the points mentioned in section 4.6 affecting the static T.E estimates as these factors still apply. Uncertainties on manufacturing errors are small though alignments are difficult to control. Tooth stiffness varies but has little effect on the end result. 3-dimensional (axial) effects should be small with low helix angles but gear body distortions and movements can have major effects.

The additional factors involved in the dynamics case are:

(i) Inertias and moments of inertia. These present no problems and are usually determined easily and accurately.

(ii) Support lateral stiffnesses and drive shaft torsional stiffnesses. These are subject to a much greater degree of error as it is difficult to assess the effective lateral stiffness of very short shafts and the bearing stiffnesses are susceptible to small changes in alignments and casing design. It is possible to measure stiffnesses in situ but bearing lateral stiffnesses and their restraining stiffness against misalignment vary with speed, load, and frequency for plain bearings and with load for rolling bearings. We conventionally make the assumption that the gearcase is rigid but all too often this is not a valid assumption. Some gearcases may deflect rather easily, reducing effective stiffness at low frequency. When above a natural frequency, the gearcase bearing housings may respond 180° out of phase. This gives motion of the housing opposing the bearing and shaft deflections and so there appears to be an increase in the support stiffness and natural frequencies are higher than expected.

(iii) Gear support damping. Damping produces more uncertainty than any other aspect of the problem as is true in most mechanical vibration engineering. It is, in general, not possible to predict it and even less possible to control it. We are dependent on experience (and possibly testing) to give a rough estimate of the damping we will get. The actual mechanism of damping in a machine is obscure since material (steel) damping is very low (typically of the level that would give resonance amplifications greater than 100), air damping is negligible and rolling bearings absorb no energy. Even plain bearings, though useful energy absorbers at once-per-revolution frequency, are far too rigid at once-per-tooth frequencies or above, so they absorb little energy. Bolted or shrink fit joints are good at absorbing energy but there are few of these in modern designs. The most effective dissipation mechanism is probably the radiation of vibration energy into the flexible casing because little of the energy returns to the rotors. A gearbox which is bolted down to the ground can dispose of much energy into its foundations but it is the energy transmitted into the supports which gives the troublesome noise in most installations. We are left with the curious deduction that an apparent improvement in the internal dynamics by altering support stiffnesses may be at the expense of radiating more energy into the structure and so increasing external noise. Lack of knowledge of support damping may not be important since damping only tends to dominate vibration response near resonances. Normally drives are kept away from resonant frequencies. If resonances can be avoided, the damping uncertainties are less important.

(iv) Tooth impact damping. This is a very important factor in determining how far apart the teeth may bounce and the frequency range over which there will be trouble. Typically we measure impact energy loss by generating an impact and determining e, the coefficient of restitution, by measuring relative velocities before and after the impact. Since this is not feasible with gears, we use the alternative approach of finding the damping while the gears are in contact from the resonant damping factor for the very high frequency modes which are associated with contact deflections. Dynamic magnification (Q) factors of the order of 10 are typical for mechanical resonances in machinery and gearboxes so we can make a good guess at damping by taking the peak damping force to be 10% of the peak elastic force during impact. Dividing by the natural frequency w_n of the contact resonance gives the damping force coefficient.

There is another uncertainty associated with damping as we tend to assume in any estimates that damping is proportional to relative velocity. The main reason for this is that all linear analysis can only deal with this assumption and estimates for hysteretic damping or more complex models of damping become rather complicated for simple analysis, whether by matrix (linear) methods or by time marching approaches. In reality the damping is probably most accurately represented by a hysteretic model but we avoid the problem to keep life simple.

In the program the damping is added with a coefficient tthdamp which is derived by taking the standard tooth stiffness coefficient $1.4 * 10^{10}$ and dividing it by a Q of 14 to give 10^9 N/m/m. Then since peak velocity is ωx if peak displacement is x, assuming a resonant frequency in contact of 1600 Hz or 10,000 rad/s, we get a damping coefficient of 10^9 N/m/m divided by 10,000 to give 10^5 N per unit velocity per unit facewidth (N s/m^2). This damping only exists if the teeth are in contact so the logic matrix (posint) which locates contact is multiplied by the relative velocity at the pitch point and the damping coefficient. The resulting force per unit length of tooth contact is turned into an equivalent elastic interference and added to the main interference to give the contact force at each slice. From an academic perspective this can be criticised because it can give slight negative values of local contact force, but the effect is very small and the alternative methods of modelling damping give much greater problems.

The main effect of uncertainties in damping is that they alter the dynamic magnification at resonances or alter the possible height of bouncing and thereby the impulsive forces and stresses. However, the frequency ranges in which trouble occurs will be little affected and it is usually where trouble happens that is of most importance, rather than exactly how high the stresses rise.

As far as estimates are concerned, all that can be done is to guess a Q (magnification) factor, as suggested above, on the basis of experience of measured values and then use this value for the estimates.

5.7 Sound predictions

The comments applicable to modelling the internal dynamics of a gearbox apply equally well to modelling the casing response. Masses and stiffnesses may be predicted with reasonable accuracy but damping is a major unknown. Unless there have been measurements on similar gearcases and installations, it is only possible to guess at Q values.

If the casing response is modelled it is possible, though laborious, to estimate the total sound power radiated from the system at the various frequencies [1,2]. Then there are the complex effects of interference between the various sound sources to generate the external sound field. Uncertainties of the order of 10 on the range of internal and casing damping factors mean that the final result is liable to be 10dB incorrect either way so the result may not be of much help as a 20 dB range is involved. It is usually more economic to follow standard design practice and then await practical tests on the casing.

Predictions for a poorly designed casing with large panels may be relatively accurate but the better the design of the casing, the more difficult it will be to make predictions. Fortunately the design rules for quiet casings are well known so it is straightforward to start with a good design.

References

1.   Lim, T.C., and Singh, R., 'A review of gear housing dynamics and acoustics literature.' NASA Contractor Report 185148 Oct 1989.
2.   Fahy, F.J. Sound and structural vibration. Academic Press, London, 1993.

6

Measurements

6.1 What to measure

As it is gearbox noise that is the problem, the obvious thing to measure is noise, with a microphone placed in typical listening positions around the installation. This, however, produces a great deal of information which is highly confused.

A microphone picks up combined noise from all the panels of a gearcase and the relatively low speed of sound in air (300 m/s compared with 5000 m/s in steel) means that at a typical tooth meshing frequency of 600 Hz the wavelength is 0.5 m. Two panels vibrating in phase 0.25 m apart will produce sound waves exactly 180° out of phase.

The interference between the waves will have a major effect on the sound and small variations of position will give major changes in sound level. In addition if there are other machines or walls near, then the reflections from the surfaces will further confuse the measurements. Fig.. 6.1 illustrates the problem.

The other effect of the speed of sound is to delay the measurement and spread it in time. If, for example, the teeth were bouncing out of contact there would be a series of impulsive waves reaching the gearcase and radiating pulses of noise. Path length differences of the order of only 0.6 m would spread the "pulses" over 2 milliseconds. A series of pulses at 500 Hz tooth frequency would then appear at a microphone as a continuous sound, making diagnosis more difficult.

The interference and reflection problems are slightly eased if we use sound intensity measurements made very close to vibrating panels. Unlike sound level measurements, sound intensity measures the amount of net sound power being transmitted in a given direction and is unaffected by reflections which may greatly increase local sound levels. Conversely, high local power emissions may be subsequently cancelled by another panel acting 180° out of phase (a dipole or the rear of a rigid body). The disadvantages lie in the high costs of the equipment and the limitation that we are just measuring the local performance of a particular resonating panel. Due to phasing effects high power radiated from one panel could be effectively cancelled by a roughly equal power radiated at the same frequency from a neighbouring panel vibrating 180° out of phase.

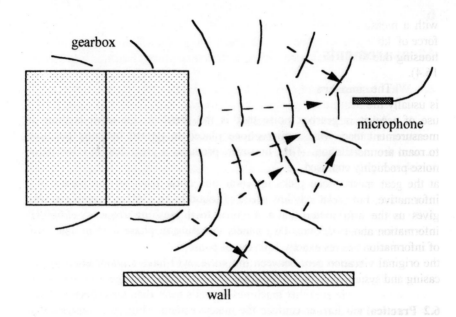

Fig 6.1 Sketch of setup indicating sound reflections and multiple paths to microphone.

The further any measurement moves away from the original source of the vibration, i.e. the contact between the teeth, the greater the opportunity for there to be vibration paths in parallel allowing complicated interactions and interferences. If we want information as uncontaminated as possible, it is desirable to go back as close to the mesh (the original source) as possible.

Measuring on the rotating shafts inside the gearbox would give us the clearest and most informative measurements but since it is experimentally difficult, this technique is only used for very special cases. Normally the first point at which we can get to the vibration is at the bearings where we would like to measure the forces coming through the bearings, but can more easily measure the housing vibration.

Housing vibration is a very simple, robust measurement using standard cheap accelerometers and it gives a good idea of the levels at the interface between the gearbox internals and the gearcase. It is then easy to use a moving coil vibrator to find the local impedances at the bearing housings so we can work backwards from the observed vibrations to determine the forces coming through the bearings. In nearly all this work the casing system is effectively linear so we can use superposition to deduce the effective exciting force at the bearing. An observed vibration of amplitude b with a measured combined local stiffness k infers an equivalent exciting force of kb. Some adjustments must be made for the vibration at one bearing housing due to the excitation forces at the other (three) bearings (see section 16.4).

The simplicity of measurement and the fact that the bearing housing is usually the nearest we can get to the trouble source, combine to make the use of accelerometers on the bearing housing the predominant method of measurement for investigating noise source problems. Using accelerometers to roam around the casing or installation allows us to deduce where the large noise-producing vibrations are occurring. Measurement of transmission error at the gear mesh, discussed in Chapter 7, is essential and is powerful and informative, but more difficult and requires more expensive equipment. It gives us the information about the excitation from the gears but not the information about the dynamic responses of the whole system. Both batches of information are needed to do a thorough investigation as the T.E. gives us the original vibration generation information and the accelerometers give the casing and system response information.

6.2 Practical measurements

As far as noise measurements and deductions are concerned there are few restrictions on measurements. Measurement of sound pressure levels is easy since a basic (digital) noise meter with analog output jack [1] can cost less than £100 ($150) and the output signal, directly proportional to sound pressure level, can go straight into an oscilloscope, a recorder or wave analyser. Direct viewing of the signal on an oscilloscope should always be used as it is very useful to get an idea of the character of a sound and whether there is a simple repetitive pattern. Synchronising the oscilloscope to once per rev of each shaft in turn gives a clear idea of whether or not there is a pattern. The alternative of using waterfall plots is sometimes less helpful especially if there is regular torque reversal during each rev as with a reciprocating engine.

At 500 Hz, a typical tooth meshing frequency, 1μm corresponds to 1g acceleration so, since we can measure down to 0.001 g with a standard piezo-electric accelerometer easily, there are no sensitivity limitations at this sort of frequency. A typical simple circuit for a charge amplifier (Fig. 6.2) gives a sensitivity of 22 mV/pC from 3 Hz to frequencies above 100 kHz. A simple fixed gain circuit works well, provided it is shielded from external electrical noise, and is extremely reliable since there are no switches or internal connections to give trouble. These advantages more than compensate for the lack of adjustment on sensitivity.

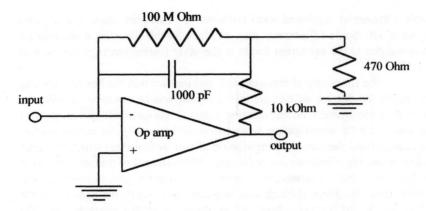

Fig 6.2 Simple circuit for fixed gain charge amplifier.

Where consistency, robustness and reliability matter, these basic single purpose circuits can be preferable to the standard commercial boxes which must cater for an extremely wide range of operating conditions and are correspondingly much more complex. A standard die cast box will take the circuit with its mains adaptor or batteries (rechargeable) and can easily be sealed against showers so that it can operate outside in all weathers.

Any high input impedance (>100MΩ) operational amplifier with a gain-times-frequency response > 1 MHz can be used. It seems wasteful but a convenient amplifier to use is an LF444 or LF347 which have 4 op-amps on a single circuit as single versions of this performance are not easily available and it is easier to use one amplifier for a range of requirements.

Using a standard [2] very economical accelerometer of mass about 20 gram, with a typical output of 27 pC/g (pico Coulombs of charge per g acceleration), we have about 600 mV per g acceleration or 60 mV per m s^{-2}. As the frequency drops, the acceleration, which is proportional to frequency squared, drops rapidly so that by 5 Hz an amplitude of 1 μm is only giving 0.0001 g and is well down into the electrical noise level unless special accelerometers are used. The electronics to deal with the small charges at low frequencies (below 1 Hz) start to become more complex. In addition, at low frequencies the equal and opposite quasi-static forces at wheel and pinion bearings tend to cancel so there is negligible vibration to measure.

None of this affects audible noise investigations since we cannot hear vibrations below 32Hz (off the bottom of the piano) unless they are incredibly powerful and they are then felt rather than heard. As mentioned previously, users who think they hear 2 or 3 Hz noise are in fact hearing modulation of much higher frequencies.

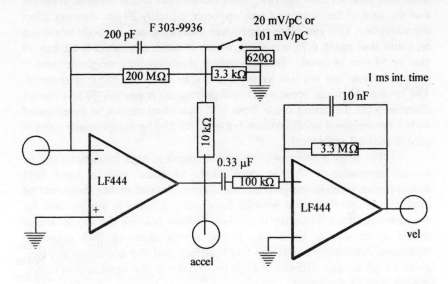

Fig 6.3 Circuit for portable vibration testing box complete with integration to velocity.

For audible noise work where the low frequencies are irrelevant the parallel resistor in the above circuit can be reduced from 100 MΩ, assisting stability of the output against sudden disturbances.

An alternative change is to use a 200 pF (1 %) capacitor in parallel with the 100 MΩ resistor to increase sensitivity allowing outputs of 100 mV per pC. Fig. 6.3 shows a circuit used for typical measurements (on a machine tool) where one stage of complication (one switch) has been added to give either 20 mV/pC or 101 mV/pC. The rolloff (3 dB) frequency at the lower end is then due to the combination of 200 pF and 100 MΩ and so is 8 Hz.

In addition, in the circuit in Fig. 6.3, another of the op-amps available on the LF444 chip has been used to give integration of the signal to velocity which is often more convenient especially as noise is proportional to velocity. The time constant is the product of the 100 kΩ and the 10 nF and so is 1 ms. This corresponds to a break frequency of 1000 rad s^{-1} which is 160 Hz so at this frequency a sine wave will be the same amplitude at output as at input. When the switch is set to the higher sensitivity the acceleration output is about 101 mV/pC x 25 pC/g or 2500 mV/g acceleration and so 250 mV per m s^{-2} and the velocity sensitivity is then 250 mV per mm s^{-1}.

At the other end of the scale, high frequencies give high accelerations and can be measured easily, but high frequencies are often associated with very low masses. The problem here is that we need to ensure that the mass of the measuring accelerometer, typically 20 gm, does not affect the vibration. This can be relevant when measuring say, car body vibrations on a thin steel panel, 0.75 mm thick, where 20 gm is equivalent to an area 50 mm by 50 mm of panel. Smaller, lighter accelerometers weighing about 5 gm can be used but are less sensitive and may still affect the measurement. The same problem can occur with small gearboxes. A gearbox 20 mm overall diameter with the casing made from 0.75 mm sheet cannot be investigated with a conventional accelerometer but may need to be exceptionally quiet if used in medical equipment.

The other problem with an accelerometer at high frequencies can be contact resonance. This is most likely to occur with a hand held accelerometer when investigating mode shapes. Pointed probes should not be used with an accelerometer because the contact stiffness is too low and the associated resonant frequency is too low. Where possible the accelerometer should be screwed or glued on. If not, a thin smear of thick grease or traditional beeswax between the (flat) surface and the accelerometer base gives a high contact stiffness at high frequencies as the squeeze film effects prevent relative movement.

If the money is available and it is necessary to measure extremely thin panels the best possible method is to use a laser Doppler vibrometer which gives velocity directly but this method is expensive and must be set carefully in position.

At one time there were problems with electronic (valve) equipment because it was necessary to have input and output impedances matched (at 600 Ω) to get maximum power transfer.

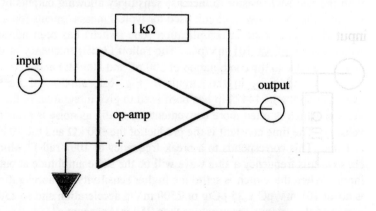

Fig 6.4 Simple current to voltage converter circuit (1 V per mA).

This is no longer a problem since most modern equipment uses voltage outputs with very low (< 2 kΩ) internal source impedance and inputs have a very high (> 1 MΩ) impedance. The exception is when long cable runs are required under electrically noisy conditions. Then a current drive may be used with a zero input impedance receiver at the far end to turn current back into voltage.

This type of amplifier is an operational amplifier with no input resistor and simply a feedback resistor to give an output voltage proportional to input current as shown in Fig. 6.4. This circuit will give 1 V per mA but only if the op-amp is capable of delivering sufficient current which is typically up to 10 or 20 mA. Alternatively it may be necessary to use a resistor of low value (10 Ω) across the inputs and then multiply the voltage as in Fig. 6.5.

Care should be taken when logging data into a computer as the multiplexing circuits may require low impedance drives to give fast settling times, so it is not possible to use simple series RC circuits on the outputs to roll off high frequency noise. The logging inputs will usually need drive impedances of less than 1 kΩ to reduce interactions between channels so that the input amplifier has time to "forget" the level of the previous channel before taking its sample. If rolloff of high frequency noise is needed it is best done by using a capacitor in parallel with the feedback resistor of the amplifier.

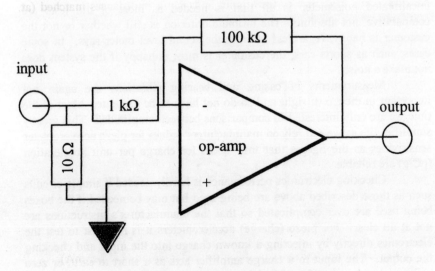

Fig 6.5 Alternative current to voltage circuit.

One method of testing internal and external resonances is to run the gearbox and use the T.E. as the excitation source, varying the speed to vary tooth frequency. The main limitation here is the inability of some gearboxes to run slowly under full torque, either because the hydrodynamic (plain) bearings will not take full load at low speed or because the gear teeth surfaces will scuff at low speed as the oil film is too thin in spite of the lower temperatures increasing the viscosity. With plain bearings there is also the problem that the shaft position alters with speed under a given load so alignments of the helices may alter as speed changes the bearing eccentricities.

As mentioned previously in section 1.6, universities, if required, can provide equipment, advice and guidance, undertake full investigations of problems, or can train personnel.

6.3 Calibrations

Calibration of instruments is in general a worry since many organisations have become enmeshed in bureaucracy and request that any measurement is traceable back to a fundamental reference.

This is a waste of time (and money) for most noise investigation and reduction work. The only time that it may be necessary to carry out an absolute measurement which is guaranteed to be accurate is if there is a legal requirement for a gearbox to be below a specified noise level. If such a test is needed then a calibrated noise meter is required but otherwise a simple uncalibrated noisemeter is all that is needed as most of the tests are comparative, not absolute. The ultimate criterion is still whether or not the customer is happy, regardless of what the sound level meter says. In some cases, such as sports cars, the customer is most unhappy if the system does not make a noise.

Measurements of casing and bearing vibrations are again not important in their own right and so do not have to be accurate. Most of the time we are only interested in comparisons between amplitudes. This greatly simplifies life as we can rely on manufacturers' values for piezo accelerometer sensitivities as the figures that they quote for charge per unit acceleration (pC/g) are reliable.

Checking electronics performance is hardly needed if simple circuits such as those described above are being used but may be needed if the boxes being used are over complicated so that the manufacturer's instructions are not at all clear. For piezo (charge) accelerometers it is simplest to test the electronics directly by injecting a known charge into the input and checking the output. The input to a charge amplifier acts as a short to earth or zero resistance as the amplifier always keeps its input at zero volts. If we have an accurate capacitor, say 100 pF and vary the voltage at input by 1 V then as the other terminal of the capacitor is held to 0 V and as q = C V there will be a charge of 100 pC injected into the charge amplifier. This gives a known input charge so we know what the amplifier output (acceleration) voltage should be.

This approach cannot be used for other types of accelerometer so unless they are the static type, which can be calibrated by turning them upside down, they are best calibrated on a vibrating table against an accelerometer with a known output.

6.4 Measurement of internal resonances

From a theoretical model (as in section 5.1) with some guesses about damping we can predict the internal responses so that we have a transfer function between relative displacement between the gear teeth (T.E.) and bearing transmitted force. Such estimates are liable to be highly inaccurate but it is almost impossible to carry out a conventional vibration response test in situ with an electromagnetic vibrator. The alternative approach is to use the tooth mesh excitation (T.E.) as the vibration source to obtain worthwhile practical results. This depends on the fact that a given pair of gears at a particular torque will have a T.E. of, say, 5 μm at once-per-tooth meshing frequency, regardless of rotation frequency.

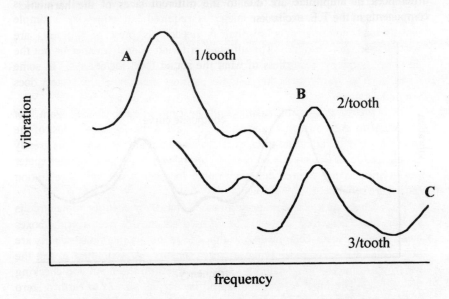

Fig 6.6 Sketch of responses to T.E. excitation as tooth frequency varies.

Varying gear drive speed (at constant torque) will give a constant relative displacement between the teeth with varying frequency and if we measure bearing housing vibration we will then have the transfer characteristic that we need between input displacement (T.E.) and output (bearing) vibration. That is, instead of sweeping a constant exciting force through a frequency range to obtain a standard resonance plot, we sweep a constant 5 µm displacement to obtain the plot.

Speed may be limited at the lower frequencies by tribology problems as in section 6.2, by the difficulty of getting high torques at low speeds on the loading dynamometers, or by the input drive motor cooling problems. At high speed the limitation is likely to be to ensure that the equipment is not oversped.

There is likely to be a 3:1 or more range of speeds possible and we have the fundamental 1/tooth component of excitation staying constant in amplitude but we are also likely to have the harmonics of tooth frequency present in the excitation. These harmonics also stay constant in amplitude provided the teeth stay in contact so that the system remains reasonably linear.

Plotting housing vibration against tooth frequency solely for the once per tooth frequency component will typically give us curve A in Fig. 6.6 and the same plot for twice tooth frequency may give curve B and thrice tooth frequency, curve C. The curves are similar where they overlap and the differences in amplitude are due to the different sizes of the harmonic components in the T.E. excitation.

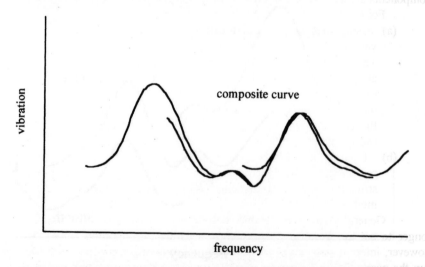

Fig 6.7 Combined curve for internal responses against harmonic frequency.

Adjusting for the variation in size allows the three curves to be collapsed into a single curve as in Fig. 6.7. This is the transfer function between T.E. and bearing housing vibration. Absolute values are only known if the sizes of the T.E components are known, but it is usually the shape of the resonances and their position relative to forcing frequencies that is of interest.

When the response is complicated with overlapping resonances it is necessary to record relative phase as well as amplitude because the phase information is valuable for identifying the resonances and separating them by the circle methods pioneered by Kennedy and Pancu [3,4].

Phase information can also be important if harmonics are being generated because it is the phase of the third harmonic relative to the fundamental which determines whether a waveform is flat topped (saturating) or peaky. Unfortunately the only reference for input phase is usually the once per revolution timing signal in a rather arbitrary position unless we have taken the trouble to set the position of the timing pulse exactly to a known (pitch point) position.

Varying speed used to present problems since only DC motors were practicable but now that three-phase inverter drives are easily available at economic prices, variable speed testing is relatively easy.

6.5 Measurement of external resonances

Measurement of the transmission path from the bearing housing vibrations to the final noise (as heard) is relatively straightforward as the components are accessible and non-rotating.

For excitation we have the choice of either:

(a) Using the gears as excitation, as with internal resonances, and varying the drive speed (using an inverter with an A.C. motor). This gives an acceleration "input" at 1/tooth, 2/tooth, 3/tooth, etc., at the bearing housings. As four or more bearing housings are excited simultaneously it is difficult to sort out the paths and determine which sources predominate. The "output" can either be the sound pressure level or the vibration level on a particular (noisy) panel.

(b) Exciting at each bearing housing in turn and measuring the responses from bearing housing to the supporting feet, surrounding structure or to a microphone. See Chapter 13 for the various methods available.

Generally (b) is preferable, despite the disadvantage that it takes longer to set up, because it is easier to separate the vibration paths. If, however, internal resonances are also being investigated it may be simpler to run the gearbox with a poor set of gears under constant torque and measure the combined internal and external resonances by measuring the bearing vibrations and the noise simultaneously. This gives T.E. to bearing vibration as well as bearing vibration to noise. Whether the bearing housing response is high or low at a resonance checks whether a given resonance is internal or external.

6.6 Isolator transmission

A gearbox will often be mounted on vibration isolators in an attempt to limit transmission of vibration away from the gearbox, e.g., in a car the combined engine and gearbox is rubber mounted to reduce vibration into the body shell.

Unfortunately isolators are often rather ineffective either because:

(a) They were designed to isolate 1/revolution (often 24.5 Hz) so they perform badly at 24/revolution (tooth frequency) due to internal resonances (spring surge) (see section 10.3); or

(b) The isolator is relatively stiff and the support flexes rather than the isolator.

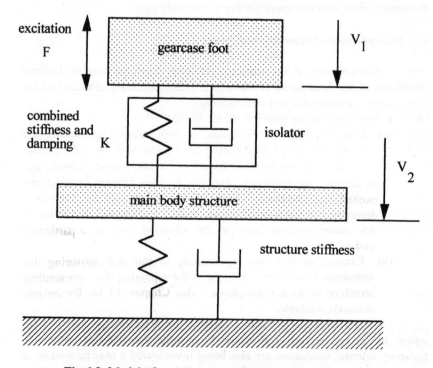

Fig 6.8 Model of an isolator in position under a gearcase.

Conventionally, it is customary to talk about the attenuation achieved by an isolator. This is measured simply by measuring the vibration above and below the isolator and taking the ratio of amplitudes.

A little thought shows that this figure is almost completely meaningless since if we mount the isolator on a massive, rigid support block there will be no vibration beneath it and the "attenuation" will be very high, regardless of the isolator characteristics whereas mounting on a very soft support will always give no attenuation through the isolator.

The isolator will have stiffness and damping and, provided it has not been designed for a frequency much lower than tooth frequency, the mass can be ignored. When the mass is negligible the response at a single frequency can be described as a ratio of amplitude of force to relative displacement with a phase lag. The supporting structure, whether car chassis, ship's hull, machine tool, etc., will also have a complex response which will involve damping, stiffness and mass with multiple resonances.

A more realistic model of the function of an isolator is shown in Fig. 6.8. There is no simple, easy test to measure the "effectiveness" of an isolator. However, it is worthwhile measuring the vibration above and below an isolator because it can give us a measure of how much vibration power is being fed into the structure via that isolator.

It is relatively easy to calibrate the dynamic stiffness (amplitude and phase) of an isolator in a separate test rig. Care is needed to get the steady component of load, the vibration amplitude and the frequency correct since isolators are often highly non-linear at small amplitudes.

Measurement of vibration above and below, taking due regard of phase, gives the relative displacement by vector subtraction and, hence, the force being transmitted by the isolator. This force, multiplied by the velocity of the supporting point gives the vibration power going into the support via that route, again taking note of phase angles.

As in Fig. 6.8, if the velocities of vibration above and below the isolator are V_1 and V_2 (complex) and the complex isolator stiffness was measured separately as K (in terms of force per unit velocity, the inverse of mobility), then

$$F = K (V_1 - V_2)$$

and the power into the hull is $F V_2$.

That part of F which is in phase with V_2 will provide the power into the main structure (and will average to half the product of the peak values, i.e., $0.5 F \times V_2 \cos \phi$). It is often easier to see what is happening by sketching out the vector (phasor) diagrams.

Isolator design is often difficult with gearboxes since reaction forces are high in relation to the weight of the gearbox. To maintain positions and alignments with high forces requires high stiffnesses whereas vibration isolation requires low stiffnesses. Occasionally highly non-linear supports may be used to alleviate this clash of requirements.

In the case of a car engine and gearbox, the supports to take the torque reaction may be spaced 1 m apart and at a full engine torque of 200 N m with 4:1 first gear ratio and 3.75:1 final drive ratio, the load on each would be 3000 N. When cruising, the load may be only 300 N (70 lbf). Ideally, to isolate 30 Hz firing frequency at idling, a natural frequency of about 10 Hz would be desirable. With an effective mass seen at a support of only about 20 kg the stiffness needed is 70 kN/m and the accelerating torque would then give 45 mm deflection, which would be excessive so a stiffening spring (or bump stop) is needed to limit travel at high torque while still isolating at low torque.

6.7 Once per revolution marker

It is a very great advantage for detailed noise investigations to have an accurate once-per-revolution marker on at least one shaft, and preferably all shafts. In the past, magnetic pickups were used but they gave a rather indeterminate waveform which varied in amplitude with speed and did not have a clear edge for accurate location regardless of speed. Standard "slotted, through scan opto-switch sensors" consist of an infra-red source and a photodetector with Schmidt trigger, and are extremely cheap so the only requirement on the rotor is a single hole, typically 1.5 mm diameter in a disc mounted on the shaft. It is not advisable to use a 60-hole disc to generate an r.p.m. count and to divide by 60 to get a once-per-revolution marker since position round the revolution is easily lost by stray pulses and averaging is then not reliable. Two separate detectors should be used if 60/rev and 1/rev are both required. An advantage of this type of marker is that its position can be set accurately, semi-statically especially if an indicator LED is fitted to show when the signal switches. Alternatively a Hall effect magnetic probe is robust and is mounted about 1 mm away from a screw head or other magnetically susceptible once per rev marker. It will give a fast acting and repeatable marker signal whose angular position does not vary with speed.

Having a 1/rev marker is an asset because:
(a) There is an exact location of any problem round the revolution especially if damage is suspected. Small scuffs and burrs can easily be located.
(b) Time averaging is reliable. This is essential if small defects are being sought in a "noisy" environment. (Here the term "noisy" does not

pertain to audible noise but is the confusing term used for any background irregular vibration, whether electrical or mechanical.)

(c) When recorded, there is an exact speed reference and when viewed on an oscilloscope the signal can be synchronised to 1/rev to give easy and rapid identification of the position of impulses or changes round the revolution. An exact speed reference also allows exact identification of whether vibration is linked to a particular shaft.

(d) An accurate 1/rev marker allows a quick check on whether a vibration is linked to 1/rev or to a harmonic of the electrical supply suggesting an electrical noise problem.

The disadvantage of a 1/rev marker is that an additional channel of information must be recorded. A slight economy of channels can be achieved by putting two once-per-rev. markers on one channel, using +ve pulses for one (pinion) channel and -ve pulses for the other (wheel) channel. Addition is by an analog operational amplifier. If combined with using pulses of different heights, this allows four markers to be identified on one channel but the pulses should be very short so that positive and negative pulses do not mask each other. At the same time the pulses must not be so short that data sampling misses some pulses as this complicates time averaging routines.

While adding/subtracting timing pulses it is also advisable to use an operational amplifier to reduce the amplitudes of the pulses (to about 1 V) and to slow down their rate of change of voltage. Standard logic TTL pulses from the sensors rise and fall 5 V in less than 0.5 μs and this sudden change gives pickup or interference between neighbouring conductors in ribbon cables or the printed circuit boards of data logging cards. Slowing down the change from rates of the order of 10^7 V/s to less than 10^4 V/s greatly reduces cross interference. This is achieved by having a capacitor in parallel with the feedback resistor on the adder, with a time constant of the order of a tenth of a millisecond or greater.

References

1. Digital sound level meter. Model 8928 obtainable from A.T.P., Tournament Way, Ashby-de-la Zouche, Leics. LE65 2UU, UK.

2. Birchall Ltd., Finchley Ave., Mildenhall, IP28 7BG. U.K. A20 accelerometers. www.djbirchall.com.

3. Kennedy, C. and Pancu, C.D.P., Journal of Aeronautical Sciences, 14, 1947. p 603.

4. Ewins, D.J., Modal testing theory and practice, Bruel & Kjaer, Harrow, 1986, Research Studies Press, Letchworth, UK.

7

Transmission Error Measurement

7.1 Original approach

Chapter 6 was concerned with the vibration and noise measurements normally made on a gearbox under operating conditions. However when problems arise we must return to the source of the vibration and measure T.E. since this is the only relevant measurement of the basic excitation that drives all the vibration. There are many possible approaches to measuring T.E. but, in practice, the use of digital encoders dominates the field.

A workable system was first developed for laboratory use by the National Engineering Laboratory at East Kilbride. It was then redesigned and developed for industrial use in the 1960s by Dr. R. G. Munro who successfully introduced the system to the Goulder (subsequently Gleason) range of gear measuring equipment. Though the objective is to measure transmission error, the check is often referred to as a single flank check [1].

Large (10") diameter rotary encoders with an accuracy of about 1 second of arc were mounted on precision spindles which also carried the meshing gears. When rotated slowly (<10 rpm), under low torque sufficient to keep the teeth in mesh, the 2 encoders each produced 2 strings of pulses (at 72 second intervals) which were processed electronically. The system is shown schematically in Fig. 7.1 with the corresponding block diagram in Fig. 7.2. The input (pinion) is driven at exactly constant speed (servo controlled) and should produce a perfectly regular string of (TTL) pulses and, with a "perfect" gear drive, the output (wheel) encoder should produce a regular string of pulses (at a different frequency).

The function of the electronics is to take the steady input pulse string and to generate the output pulse string expected if the gear drive were "perfect." The "perfect" string is then compared with the real, measured string and any variation in phase angle between the two strings corresponds to an angular error in the drive. The requirement for a servo-controlled steady input speed is due to the requirement for multiplying the input frequency by a ratio corresponding to the number of teeth (W) on the output wheel. The phase-lock loop which achieves this cannot deal accurately with the rapid variations in frequency which would occur with torsional vibration at the input.

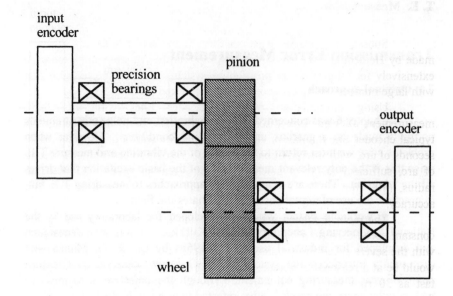

Fig 7.1 Sketch of setup of Goulder type single flank tester.

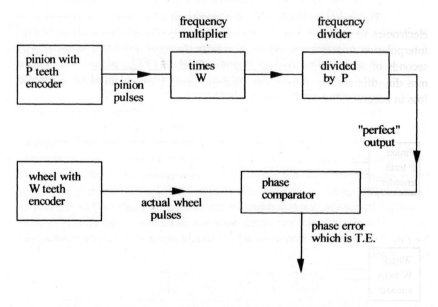

Fig 7.2 Block diagram of original (single flank) T.E tester.

Subsequent designs used smaller, more robust encoders, usually made by Heidenhain [2] which had become readily available and were used extensively for (static) rotary positioning systems on machine tools so that with large numbers being produced they were priced economically.

Using interpolation between encoder lines with the phase measurement allowed finer discrimination than the basic line spacing. A typical encoder line number of 18,000 lines per rev with a line spacing of 72 seconds of arc, with interpolation, could easily resolve to better than 1 second of arc, sufficient for most machine tool and gear purposes. At 200 mm radius, (400 mm or 16" diameter) 1 second of arc corresponds to 1 μm accuracy.

The main problems with this approach lay with the need for a very constant speed drive to allow the frequency multipliers to work correctly and with the severe speed limitations. The original pulse strings from the pinion would be at a reasonable frequency of 9,000 Hz if the pinion was rotating as fast as 30 rpm but if there were 106 teeth on the wheel the multiplied frequency would be 954 kHz, which was faster than the available electronics could handle comfortably.

7.2 Batching approach

The next approach uses the same encoders but uses interpolation electronics to generate many more pulses per rev. Typically there is 50-fold interpolation so that an encoder with 18,000 lines per rev gives pulses at 0.36 seconds of arc (0.0001°) spacing. This gives a fine resolution since on 100 mm dia. this corresponds to less than 0.1 μm. There is however possibly a loss in accuracy compared with the original signals direct from the encoders.

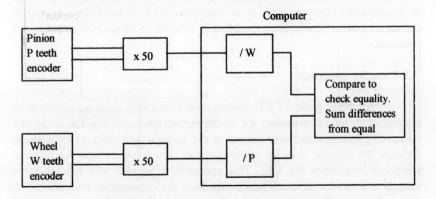

Fig 7.3 Block diagram for batching approach.

The system is computer-based and counts the interpolated pulses from the pinion and the wheel encoders and compares the expected number of pulses from the wheel with the observed number. This approach avoids the need for multiplying phase-lock loops so there is no longer such a critical requirement for constant input speed. The necessary interpolating interface cards can be obtained directly from Heidenhain Ltd. [2].

In a typical case of a 21 to 106 reduction drive the computer could count 21 pulses from the wheel pulse string and determine how many pulses there were in that time from the pinion encoder. The correct number of pulses, 106, would mean that during that time interval, (corresponding to about half a minute of arc rotation of the pinion) there was no change in the value of the transmission error. Any variation from the expected number would raise or lower the T.E. by increments of 0.36 sec arc. Fig. 7.3 shows a block diagram of the principle. The diagram is similar to the original system but after the initial multiplication by 50 (instead of multiplying and dividing on one string) both strings are divided.

This interpolation system works well but again suffers from a fundamental speed limitation. If interpolation is to 0.36 second of arc then there are 360 x 60 x 60 / 0.36 pulses per rev or 3,600,000 pulses per rev. Typically, electronic systems use 0.5 microsecond TTL pulses so, for reliability, the frequencies should not exceed 1 MHz and rotation speeds are then limited to about 0.25 rev/s or 15 rpm. This is perfectly satisfactory for inspection purposes but not for test and development. These frequencies are sufficiently high to prevent simple programming of a PC for on-line use as the computer is not happy if asked simultaneously to accept data, calculate the result and output data. The alternative is either to record the pulse strings or to use one pulse string to gate the other then process the information off line. Both give lower speeds. Working off-line is perfectly satisfactory for research or development purposes but may be restrictive for high production or for test bed development where time available is limited so immediate answers are required.

7.3 Velocity approach

A further group of T.E. systems work on a very different approach as instead of using the encoders for direct measurement of angular errors the velocity approach effectively measures the angular velocities of each shaft, deduces the angular velocity vibrations then integrates to find the angular vibrations and hence the T.E. This approach is popular with researchers as though it is slow it is much less costly than the commercial equipment. A relatively coarse line spacing is used on the encoders and a high frequency timer (100 MHz) in the computer measures the time between encoder pulses.

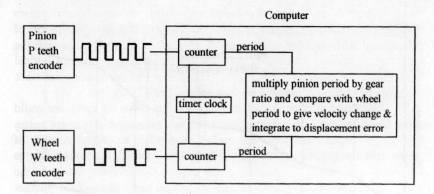

Fig 7.4 Simple block diagram of one velocity approach.

The system then effectively calculates instantaneous speeds for each gear separately, subtracts the correct (average) speed of that gear then integrates speed errors to get angular errors so that angular vibration for each gear is determined. Input angular vibration is then scaled by the gear ratio to get the expected output angular vibration and this is subtracted from the observed output angular vibration to get the T.E.

An alternative view of these methods is that (instead of as in section 7.2 one encoder pulse string being used to gate the other) each encoder pulse string is used to gate a high frequency timing signal. There are several variations possible on this theme and alternatively each encoder pulse train may be demodulated in the computer to extract the torsional vibration. Sweeney and Randall [3] and Remond [4] describe different processing methods though some of their comments about the disadvantages of alternative methods are not correct. Tuma [5] has a similar system which again takes the original pulse strings and demodulates them to determine vibration on each gear separately. Fig. 7.4 shows the simplest block diagram for a velocity system.

The principle is shown diagrammatically in Fig. 7.5 which plots angular displacement against time. For each encoder the pulses come at roughly equal time intervals and measuring the interval exactly gives the velocity which is the slope of the curve. The calculated velocities of the input can be adjusted by the velocity ratio to give the expected velocities and displacements at output and plotted on the same graph as the measured output displacements. Some interpolation is required to give the difference between expected and observed output displacements which is the required T.E.

As this approach is measuring speed variations it also gives the local torsional vibrations of the individual gears which may be of use for research if speed variations are being matched to a computer model.

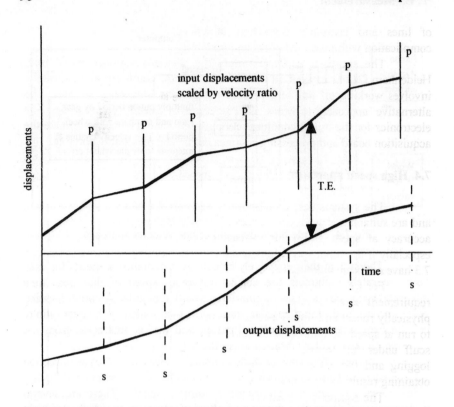

Fig 7.5. Sketch of velocity approach principle. Points p correspond to positions of input encoder pulses and the slope between two samples is given by the time interval. The slope is adjusted by the velocity ratio. Points s are where the output encoder pulses rise and again the slopes are given by the timer.

The velocity method has few speed restraints since if we have 5000 pulses per rev and a typical computer speed limitation of 100 kHz input the speed can rise to 1200 rpm. This is less of an advantage than it might appear as vibration information above 600 Hz is very likely to be distorted by system resonances and this limits us to 300 rpm for 5 th harmonic of 24 teeth.

In theory a low line spacing can be used because if we take the rough rule of thumb that, for easy visualisation, we want about six data points to locate a sine cycle then for 100 teeth and information up to the 5th harmonic we need 3000 data points per revolution (of the slower gear, i.e., the wheel). This means that an encoder with as few pulses as 3600 / rev could be used. However in practice accurate encoders are rarely available with low numbers of lines and computer correction of encoder errors is an unwanted complication with increased effort and possibility of errors.

The approach can use simply the standard interface board from Heidenhain [2] to carry out all the processing in the computer but this usually involves working off line and so does not give an immediate answer. The alternative approach to work in real time at speed requires specialist electronics for the initial counting and buffering, and a computer to take the acquisition board and processing routines but can be reasonably portable.

7.4 High speed approach

The systems described above in sections 7.1 and 7.2 will work well and are suitable for production checking provided there is no requirement for accuracy at speed since the systems cannot provide accuracy at speed, especially if the input speed is fluctuating. The systems described in section 7.3 have not been widely used.

For troubleshooting, development and consultancy work there was a requirement about 20 years ago for equipment which was very compact, physically robust and highly portable to take to test "in situ" with the ability to run at speed, so that bearings could operate correctly and teeth would not scuff under full torque. As with much urgent development work, data logging and the associated computing were not necessary when speed of obtaining results was the priority.

The equipment developed (at Cambridge) would fit easily into cabin hand luggage for flying and uses two medium-sized (100 mm, 4", diameter) encoders which could operate up to 6000 rpm. Accuracy of the encoders, made by Heidenhain [2], is usually about 2 seconds of arc peak to peak, more than sufficient to meet the requirement for noise investigations because the encoder accuracy at frequencies greater than 20 times per revolution is better than 0.1 seconds of arc [6].

The electronic system used in practice is an extremely simple and robust "double-divide" system with the block diagram shown in Fig. 7.6. Complicated variations with extra multipliers (which are temperamental) could be used [7] but it has not been found necessary for any normal drives. As usual, the encoders are mounted on support flanges attached to the gear casing and are driven from the free ends of the gear shafts via flexible connectors which allow for slight misalignment but are extremely stiff torsionally to keep the torsional natural frequency high. Alternatively the encoders can be the shaft mounted variety if there is a sufficiently robust shaft extension to support them.

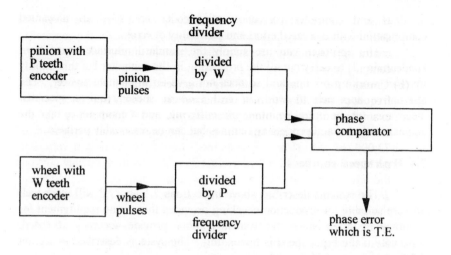

Fig 7.6 Block diagram of high speed T.E. tester.

Typically, with a gear ratio of 19:31 and the standard 18,000 (x 4) line encoders and 12 bit digital recording, the resolution would be to about 0.1 second of arc and more than adequate for automobile work. The detailed design of the electronics requires some care to get the high accuracy necessary for the final phase comparison.

This approach, unlike that in section 7.1 (the original pulse frequency multiplying system) is not affected by torsional vibrations at the input so it can be used under industrial conditions in situ on machinery such as printing machines. The encoders can be mounted quite large distances apart (50 m) on printing rolls to investigate dot synchronisation problems and the results show clearly when the gears start coming out of contact.

Although intended originally for on site work at moderate speeds of the order of 200 rpm there seems to be no practical limit to operating speeds other than the requirement that for the simplest and most robust system the vibration should not be so severe that there is reversal of rotation. This is a requirement for most of the systems.

Practical limitations of the "double-divide" high speed system arise from three sources:

(a) Since the system will operate over very large frequency ranges from roughly 0.01 rpm to 6000 rpm, the operator needs to dial tooth numbers, roughly to zero the trace on the screen and to set a low-pass filter according to the conditions so the system is not completely automatic and "idiot proof." This makes it suitable for development or consultancy work but less suitable for production monitoring using unskilled personnel when speeds and tooth numbers change frequently.

If on the other hand a test rig is set up to test a particular drive, the settings remain constant and the only requirement is for the operator to centralise the trace on the monitor. This can alternatively be done by computer control.

(b) Unusual tooth numbers with large numbers of teeth give too coarse a frequency resolution to pick out harmonics of tooth frequency. If, for example, a 68 tooth pinion is meshing with a 313 tooth wheel then the carrier wave which contains the phase information is at a frequency of 72,000/(68 x 313) or 3.3 times tooth frequency. Setting a very high performance filter (8-th order elliptic) to 2 x tooth frequency cuts out the unwanted "carrier" frequency but means that only the 1/tooth and 2/tooth components of error can be measured. This is rarely a limitation. It can be avoided by using multiplying circuits [7] as in the original system but measurement is then influenced by vibration. Results can also be obtained by using an approximate ratio, see section 7.10.

(c) Encoder dynamics. The encoders are driven via light but torsionally stiff couplings, but it is not possible to get the torsional resonant frequency much above 1500 Hz so the useful operating range is limited to about 1200 Hz even with the most careful design of the coupling. To achieve this performance there must be an accessible free end on each shaft. The alternative is using encoders mounted directly on the drive shafts but this gives only a limited improvement in frequency range.

(d) Non-synchronous drives. Occasionally it is not possible to mount an encoder directly on a gearshaft so a friction or belt drive is used. This tends to limit the rig dynamics severely and also the drive is no longer an exact ratio. The drive ratio can usually be approximated with sufficient accuracy using tooth numbers of less than 100.

An alternative variant possible for on-line high speed work is a blend of the high speed and velocity approaches. Each encoder string is processed separately and is simply taken to a demodulator. The two resulting vibrations can then either be logged separately or the input can be scaled by the velocity ratio and subtracted from the output to give the T.E. This method appears simple but costs increase as it requires two expensive filters (demodulators) and two accurate flip-flops rather than one and there is an additional scaling involved with possible errors due to small differences of two large quantities.

There are different approaches to demodulating the pulse string from an encoder. One used by Tuma [5] involves the analytic extraction of the phase of the pulse string and the steady increase of the phase corresponds to the rotational speed while the variations correspond to the torsional vibration. At each transition from $+\pi/2$ to $-\pi/2$ the analysis needs to add the value π.

This method is difficult to implement in real time so is more suitable for research where time scales do not need to be short.

The corresponding analog approach uses a phase locked loop to generate a reference pulse string at the average rotation speed. The phase locked loop behaves as a seismic system with a second order characteristic and will give good vibration information above the natural frequency of the loop while ignoring speed variations or vibrations well below the natural frequency of the loop. Fig. 7.7 shows a block diagram for one loop but two are needed for T.E. determination. The divider can be set to any number that allows the output not to exceed full scale but setting the dividers on the two channels to numbers which approximate to the gear drive ratio simplifies subsequent subtraction.

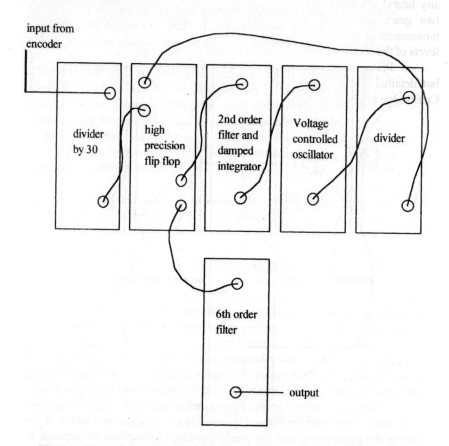

Fig 7.7 Block diagram for phase-locked loop for one encoder channel.

The natural frequency of the loop can be set very low so that the output still records the once per revolution conponents of torsional vibration but it is more customary to set the loop frequency at about a third of tooth frequency. The output then gives the tooth frequency and harmonics which are relevant for noise investigations while ignoring eccentricities which are not of interest for noise.

7.5 Tangential accelerometers

One alternative analog method of measuring T.E. is by the use of tangentially mounted accelerometers to measure the torsional accelerations of each of the shafts as sketched in Fig. 7.8. Two matched accelerometers are used and their outputs are summed into the single charge amplifier so that any lateral vibrations are self cancelling. The torsional accelerations of the two gears are scaled, proportional to the diameters, to give tangential movements at the pitch radii and subtracted to leave the T.E. En route, the levels of the torsional vibration in the system are obtained.

Previous attempts using this approach had achieved limited success but detailed checks against encoder measurement of torsional vibration at Cambridge established that:

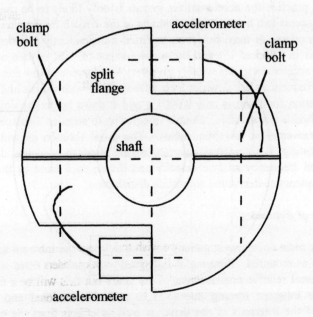

Fig 7.8 Torsional accelerometer arrangement.

(a) Information at 1/rev was not reliable and should be discarded. In operation the low cut frequency on the charge amplifiers was set to attenuate the 1/rev part of the signal but to pass tooth frequency.

(b) There was good agreement between accelerometers and encoders in the middle frequency range.

(c) The accelerometers appeared to give reliable information at high frequencies (>1 kHz) where the encoders were no longer reliable due to torsional resonances.

The advantage of the accelerometer system is the extended frequency range at the upper end and the relative ease of fitting tangential accelerometers with a clamped flange compared with aligning encoders and using delicate high frequency couplings. The accelerometers do not need a free shaft end.

The flange needs care as the match to the shaft should be good, the flange should be light and the clamping powerful enough to ensure that the accelerometers follow the shaft vibrations faithfully.

Corresponding disadvantages are the 1/rev spurious results due to gravity interacting with accelerometer axis misalignment and the major problems of supplying electrical power and buffering out the signal on a rotating system as slip rings or telemetry tend to be expensive or temperamental.

In practice the accelerometer system is only likely to be used when there is no access to a free end of both shafts or the 1/tooth frequencies are too high for encoders. It may, however, be fitted independently for monitoring purposes (as in Chapter 15) and the measurement of T.E. is then a bonus. Tangential accelerometers inevitably give very low outputs at low frequencies so if tooth frequencies are down at 5 Hz as may occur with worms and wheels the acceleration for 1 μm is only 0.0001 g and is down below the noise level so this method is not suitable. Double integration to angular displacement is also temperamental at low frequencies. The usual solution as with much vibration testing is to analog integrate acceleration to velocity, data log velocity then frequency analyse velocity and divide each band by the mean angular frequency to derive the amplitude distribution.

7.6 Effects of dynamics

For most noise investigations we wish to know the inherent accuracy of the gears as mounted. Running at full speed with encoders fitted will give us the torsional relative displacement of the gears but this will be a function of both the inherent forcing due to T.E. and the torsional and lateral vibrations of the internals of the drive as well as effects from the external drive system dynamics.

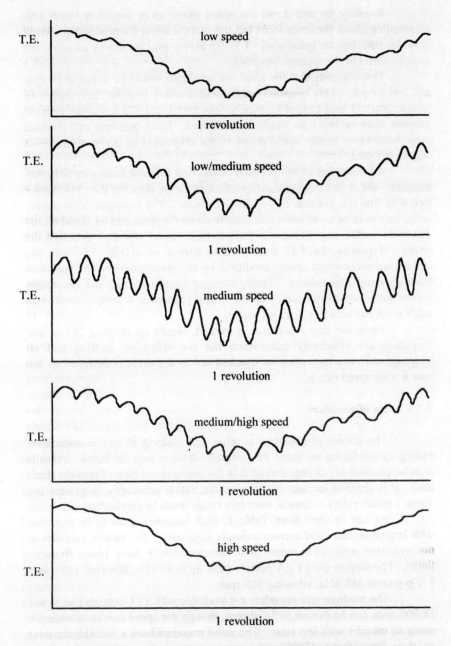

T.E.   low speed

                                        1 revolution

T.E.   low/medium speed

                                        1 revolution

T.E.   medium speed

                                        1 revolution

T.E.   medium/high speed

                                        1 revolution

T.E.   high speed

                                        1 revolution

Fig 7.9 Variation of T.E. with speed due to internal dynamics.

Running up and down the speed range as in section 6.4 will give information about the resonances but the main interest for production control is in determining the quasi-static T.E. (to assess gear accuracy) avoiding the complications of the system dynamics.

This suggests that the ideal test condition would be to run at 10 rpm and full torque. This is usually not possible either because drive motor or (dynamometer) load cannot operate at low speed and full torque or because gearbox teeth or bearings would be destroyed. Plain bearings will increase their eccentricity as the speed drops so the alignment of the meshing gears may be affected.

A knowledge of the position of the first internal resonance is highly desirable, either from theoretical predictions or by running the drive under torque to find the position of the first resonance. The resonance may appear either as a peak or as an anti-resonance because the measured torsional effects due to the mesh may decrease if there is high lateral vibration to absorb the errors. Typically the T.E. traces would appear as in Fig. 7.9, with the underlying eccentricity effects unaltered by the speed changes but once-per-tooth showing a resonance. Tooth meshing conditions may not be exactly correct but since the frequency of the lowest resonance is very insensitive to tooth mesh stiffness this does not matter.

Once the first resonance is located, results up to about 2/3 of that frequency are effectively quasi-static but the effect on scuffing and on hydrodynamic bearings must be checked unless the drive is designed to run over a wide speed range.

7.7 Choice of encoders

The choice of encoders is wide and looking at any manufacturer's catalog is confusing as some 50 different designs may be listed. Absolute angular position is not required so it is the incremental type of encoder that is used. It is simplest to classify the encoders, rather arbitrarily, in groups as in Table 1 which refers to typical sizes in a range made by Heidenhain [2].

As can be seen from Table 1, high accuracy tends to be associated with large diameter (and correspondingly high cost). The largest encoders are not available with TTL output and correspondingly have lower frequency limits. The outputs are 11 μA peak to peak up to 90 kHz, allowing 150 rpm or 1 V p-p up to 180 kHz, allowing 300 rpm.

The medium size encoders are available with TTL outputs and so with 18,000 lines can be run up to 3330 rpm, though the speed can be increased by using an encoder with less lines. The small encoders have a 300 kHz limit but, as they have fewer (5000) lines, can operate up to 3600 rpm before encountering the frequency limitation.

Table 1 - Encoder parameters

Dia. mm Mass kg	Shaft type	Accuracy Sec arc	No of lines typically	Output	Name
170 2.8	Solid 14 φ	±1	36000	11 μA 90 kHz	ROD 800
200 3.3	60 mm bore	±1	36000	1 V 180kHz	RON 886
110 0.7	Solid 10 φ	±5	18000	TTL 1 MHz	ROD 260
110 0.8	20 mm bore	±5	18000	TTL 1 MHz	ROD 225
58 0.25	12 mm bore	±13	5000	TTL 300 kHz	ERN 420
58 0.25	Solid 10 φ	±13	5000	TTL 300 kHz	ROD 420
36.5 0.1	Solid 4 φ	±18	3600	TTL 300 kHz	ROD 1020
36.5 0.1	6 mm bore	±18	3600	TTL 300 kHz	ERN 1020

Encoder price is roughly proportional to weight so there is a financial incentive to use the smaller encoders. All encoders have axial length less than 50 mm.

When mounting encoders onto a gearbox, choosing between a through-bore or stub shaft installation can be difficult. If there is a through shaft such as a collet operating rod then there is no choice and an encoder with sufficient through bore must be used. Otherwise, with a free shaft end, the choice is complex but is dictated by the mechanics of the test setup.

The through-bore type is usually completely supported by the gear shaft extension and so the installation is simple with high torsional natural frequencies, typically above 1000 Hz even for the medium-sized encoders. Reference to "earth" requires a restraint arm as long as possible with rigid light construction so there are small angular movements of the stator due to any eccentricities. The corresponding disadvantages are that the shaft must run true or lateral vibrations will be high and the shaft must be strong enough to take the weight and vibration of the encoder body. This is not usually true if an extension has been bonded or pressed onto an existing (short) gearshaft. The overhung mass of the encoder may be large in relation to gear masses and so may give an extra low frequency resonance.

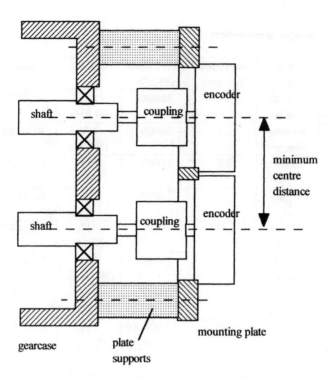

Fig 7.10 Encoder mounting at shaft ends.

Installation of the stub shaft type of encoder is more difficult as the main body has to be held by bolting onto a mounting plate which is itself supported off the end face of the gearbox as in Fig. 7.10 The plate should be mounted sufficiently accurately to ensure that the encoder is aligned to the gearshaft extension within about 25 μm and the gearshaft extension should be running true within about 25 μm so that the flexible coupling between them does not have to cope with large misalignments.   The manufacturers can supply suitable couplings (such as the KO3) which are torsionally very rigid to maintain high torsional natural frequencies but are flexible laterally as the encoders must not be subjected to high (10 N) spindle loads either axially or laterally. The mounting plates for the encoders must be mounted very rigidly to the end face of the gearcase since if they vibrate torsionally the information will not be correct.

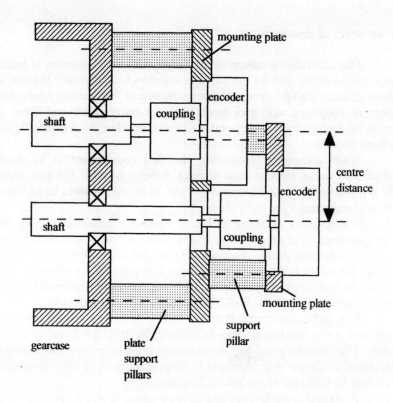

Fig 7.11 Staggered mounting with low centre distance.

A complication can arise with either through or stub mounting since the centre distance of the gear pair may not accommodate the two encoder radii and one shaft must be extended to allow the encoders (and if necessary their couplings) to be staggered axially as in Fig. 7.11. Use of smaller encoders such as the 58 mm diameter encoders helps greatly as the centre distance can then be 60 mm without stagger or about 40 mm with an extended shaft and stagger.

The smallest practical size is 36.5 mm diameter and without stagger the centre distance is 37 mm or with maximum stagger the centre distance can be about 22 mm. Unfortunately this involves having a shaft extension which is long and slender, making it difficult to ensure that it runs true. Long shaft extensions make it more likely that gearbox dynamics will be altered if the encoder is shaft mounted or that the flexible coupling has to accommodate large eccentricities.

7.8 Accuracy of measurement

The calibrated accuracy of the larger (150 mm) encoders is better than 1 second of arc and for the 100 mm encoders used normally is about 2 seconds of arc. Careful design and manufacture of the necessary torsional diaphragm couplings will give errors that are undetectable and there is virtually no limit to the accuracy obtainable with the electronics especially for low tooth numbers.

When comparing accuracies, the first requirement is to check whether peak value, peak to peak or r.m.s. is being quoted. For gear noise work it is p-p which is most frequently of use, so on the encoders listed above there is a range from 2 sec to 36 sec arc.

The resulting accuracy of T.E. is controlled predominantly by the encoder accuracy. For drives which need absolute accuracy, such as printing drives or positioning drives, the quoted ±1 second (ROD 800) or ±2 to ±5 seconds (ROD 260) is the relevant accuracy. It is possible to improve on this accuracy by first using a dynamic back-to-back calibration technique which gives the individual errors at, say, 2000 points round an encoder [8].

This information can then be used to computer-correct observed results and get a significant gain in accuracy so that ± 0.1 sec of arc is feasible. For noise purposes, this is not needed since we get a major accuracy bonus because we are only interested in frequencies such as tooth frequency which is at 15 times per rev or greater frequencies.

A typical manufacturer's calibration curve is shown in Fig.. 7.12 and the major components of error are of frequency less than 5 times per rev and at line frequency (18000 times/rev) or greater. Initial calibration checks on the large encoders gave errors of less than 0.03 sec arc at 15/rev harmonics and above and subsequent tests on medium size encoders (ROD220) also showed errors well under 0.1 sec [6].

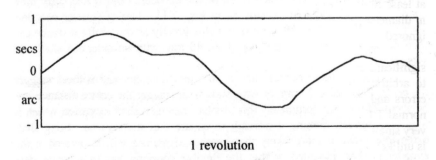

1 revolution

Fig 7.12 Typical error curve for an encoder.

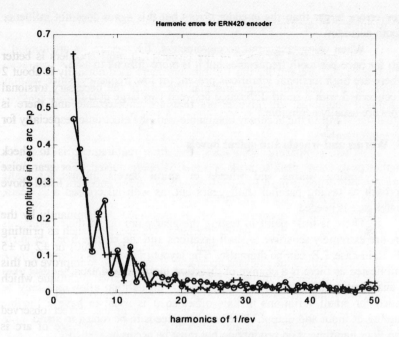

Fig 7.13 Frequency analysis of encoder position errors.

Recently, tests were carried out on the small size of encoder (ERN420) with a nominal accuracy of 26 sec arc p-p for the 5000 line version. The results were very encouraging as the errors for components at frequencies above 15/rev were well below 0.1 sec and were consistent to well within this figure. Fig. 7.13 shows results for the frequency analysis of the errors for 2 test runs in the same direction. As errors at tooth frequencies are at least 30 dB down and are less than 0.03 of a second of arc then even on 1 m diameter gear this corresponds to less than a tenth of a micron and may be ignored.

The error curves supplied by the manufacturers may sometimes show significant errors at frequencies such as 98/rev but these false errors are due to arbitrary sampling techniques which pick up and alias high frequency errors and do not necessarily appear when the encoders are being used for normal T.E. measurement especially when checking worms and wheels. The very small encoders are less accurate but accuracy at once per tooth frequency is unlikely to be a problem since the radii of the gears are so small. Typically with a gear only 50 mm diameter, as 1 sec arc is 4.85 μradian, 0.5 μm error is 4 sec arc. When using the very small encoders the coupling is liable to give

1/rev errors larger than the encoder errors but this again does not influence 1/tooth accuracy.

When using tangential accelerometers, T.E. accuracies are normally high for once-per-tooth frequencies but it is more difficult to assess accuracies if there are high torsional vibrations present (at low frequency) since we may be concerned with a small difference between two large quantities. However, errors are usually negligible.

7.9 Worms and wheels and spiral bevels

Testing worms and wheels or spiral bevels follows the same approach as testing parallel shaft gears but, as with all crossed axis gears, greater care is needed.

There is little point in testing the gear pairs out of their casings as they are extremely sensitive to shaft positions and the change from setup rig T.E. to in-case T.E. can be dramatic. The layout of the test is inevitably more complicated as there is a change of direction involved and usually offset axes so auxiliary packing blocks must be made. In addition allowance may be needed for small variations in axis offset so it is usual to have a flexible coupling at input and output. The coupling needs to be robust to stand up to shop floor handling if on production but must be accurate.

Worms and wheels have the complication that the critical once per tooth frequency is at once per rev of the input so coupling and encoder should be reasonably accurate. Designs which have an internal coupling between a motor and worm are difficult to test in the completed state as errors which are at "once per tooth" or harmonics can be gear or coupling. When the drive is being used for positioning and accuracy is important it is advisable to have some system such as double eccentrics for varying the position and directions of the worm axis relative to the wheel to allow selection of the best meshing conditions to minimise the 1/tooth component.

When testing accurate worms and wheels intended for positioning use in the metrology lab it is advisable to run the input rather faster than normal since an input speed of 10 rpm and a reduction ratio of 360 to 1 would involve a wait of 36 minutes for each output rev. This suggests that an input speed of about 200 rpm would be more suitable and so the tendency is to use a smaller encoder at input especially as high accuracy is not needed. The smallest size of encoder is not usually suitable as the bellows type of coupling for 4 mm dia is not accurate at 1/rev. There should also be an integral ratio between the numbers of lines at input and output to simplify the setting of the ratios.

Similar considerations apply to spiral bevel drives but they are usually used for high powers rather than accuracy so there is liable to be heat generation in situ. This can mean that to get realistic T.E. results in a metrology lab, it is necessary to preheat a rear axle differential unit to about 70° C to get results which are representative of in-service conditions. This is especially relevant for aluminium alloy casings.

7.10 Practical problems

An extension of the use of T.E. (single flank) checking is to measure the errors on one (drive) flank then, without altering settings or losing position, to transfer to the "back" flank and measure that. The resultant plot gives not only the errors but the variation in backlash which may be crucial for control drives or very accurate positioning systems. The pulse processing is in general more complex as it must account for direction changes if drive direction is reversed but if it is possible to reverse the load torque to transfer to the other flank while continuing to rotate in the same direction this does not involve change of rotation direction and so all the systems will cope.

The encoder systems, other than the batching approach, rely on the basic assumption that between sampling pulses there are negligible variations in speed. In practice this is true unless a ridiculously low number of encoder lines is used for the velocity approach or there are very high tooth numbers with the high speed approach.

Testing parallel shaft gears as pairs in their unmounted state allows extra testing to be done to check the effects of misalignment on the mesh.

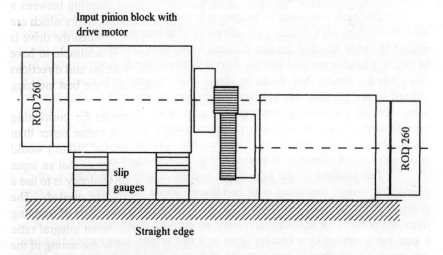

Fig 7.14 Diagram of plan view of setup on surface table for parallel axis checking.

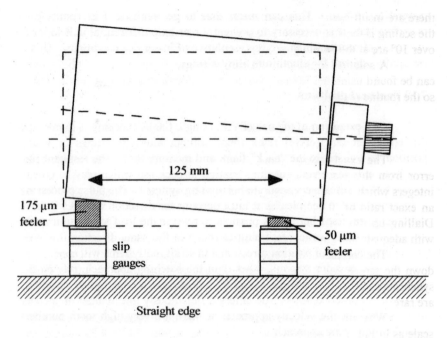

125 mm

175 µm
feeler

slip
gauges

50 µm
feeler

Straight edge

Fig 7.15 Use of two feelers to prevent centre distance variation when checking misalignment.

The basic setup can be as shown in plan view in Fig. 7.14 where an accurate straight edge is used as a reference. One bearing block is positioned against the edge and slip gauges are used to position the other bearing block so that it is exactly parallel and the correct centre distance away. Testing like this gives the results that would be obtained if the gear axes were perfectly aligned in the gearbox but it is sometimes very worthwhile in development being able to deliberately misalign the gear axes to check sensitivity to manufacturing errors or deflections. Feeler gauges may be used or the two stacks of slip gauges altered but it is advisable to ensure that centre distance at the gears is not altered as indicated in Fig. 7.15, exaggerated.

The problem in the high speed system of very large tooth numbers giving too coarse sampling was mentioned above and there is a linked problem if for unusual reasons the drive is not exactly synchronous. The latter can occur if for operational reasons an encoder is not directly coupled to a gear but is driven by a friction drive or a belt drive. Large tooth numbers can occur if a gearbox is two-stage as with, say, 19:27 first mesh and 31:34 second mesh, the overall ratio is 589:918 with no common factors. If lack of space involves using small encoders which only give 20,000 pulses per rev, there are insufficient pulses to allow measurement of 1/tooth frequencies and the scaling is too coarse as full scale would be 360 x 60 x 60 x 589 / 20,000 or over 10° arc at the output gear.

A solution to the problem can be to use an approximate ratio which can be found using the Matlab routine 'rat'. The ratio in this case is 0.64161 so the routine reads

[N,D] = rat(0.64161, 0.0005)

The exact ratio is input together with a figure for the permissible error from the exact value and the routine returns the values of the lowest integers which will approximate the ratio. The routine returns 34:53 which is an exact ratio of 0.64151 and so only 0.0001 away from the correct value. Dialling up this will allow measurement to a sensible full scale value and with adequate margin between 1/tooth and carrier frequencies.

The corresponding penalty is that the trace will gradually creep up or down the screen and exceed the limits, reappearing at the other limit. Non-synchronous ratios between 0.99 and 1.01 present problems but these ratios are rare.

When the system is run, the output would normally appear either on scale as in Fig. 7.16 or going over the limits as in Fig. 7.17. In the latter case the trace is brought into range by injecting pulses into one or other encoder string until the trace is roughly central as in Fig. 7.16.

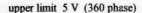
upper limit 5 V (360 phase)

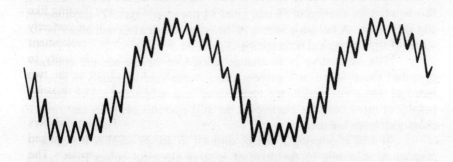

lower limit -5 V (0 phase)

Fig 7.16   T.E. trace centred on screen.

upper limit 5 V (360 phase)

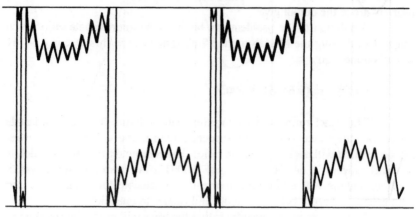

lower limit -5 V (0 phase)

Fig 7.17 T.E. trace exceeding limits.

With a non-exact ratio or with microslip at a drive joint or with a friction or belt non-synchronous drive the trace will drift as indicated in Fig. 7.18. This is of course a nuisance but provided that the trace does not drift out of range within, say, 4 revs of the input it is possible to record 4 revs and the resulting frequency analysis will be sufficiently accurate. Taking a value for drift which is 0.0003 away from the correct value means that each revolution the drift will be 0.0003 of 360° which is 389 sec of arc. Turning this into μm for a radius of 50 mm gives 94 μm so in 4 revs 377 μm apparent slip will occur. A full scale setting of the order of twice this will allow for the 4 revs of slip and typical eccentricities.

The alternative is to change allegiance to the seismic approach described above where each encoder string is analysed separately to give the torsional vibrations which are then scaled and subtracted. This has the penalty of more complex electronics but still operates easily in real time if phase lock loops are used.

If drift is ocurring it can be difficult to decide whether the cause is mechanical microslip in the drive or is stray electrical pulses from mains interference. Variation of torque may solve the problem or if count activity occurs on the dividers of the high speed system when the rig is stationary. Operating an electric drill which is plugged in to a neighbouring socket may induce a response if a system is noise spike sensitive.

upper limit 5 V (360 phase)

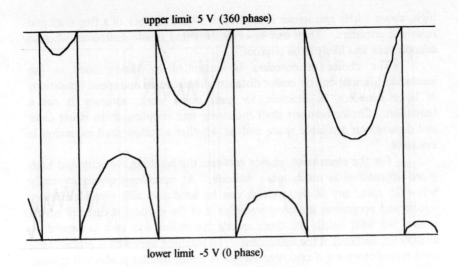

lower limit -5 V (0 phase)

Fig 7.18 T.E. trace with drift occurring.

This problem of noise sensitivity can be greatly reduced in a specific case by altering the interface electronics so that the encoder signal comparators at input have a slow switching response suitable for the particular (slow) test conditions. The system interfaces must then be altered back to normal if high speed tests are subsequently required.

In practice it can very be difficult to prevent microslip if rigid couplings are used when testing gears in situ in a gearbox so it is very advisable to use flexible couplings such as the Heidenhain KO3 which is designed to give both accurate drive and high torsional rigidity to keep natural frequencies high. Correspondingly any support system for an encoder body or torsional restraint system requires care to prevent vibration.

7.11 Comparisons

With five differing methods of measuring T.E. available it is, at first sight, rather difficult to make a choice. However the original approach is now no longer used due to the restrictions imposed by the multipliers so it can be ruled out.

Tangential accelerometers will not give useful results at once per rev or at low frequencies and so are unlikely to be used in a metrology laboratory where test speeds are very low. Running at speed would give tooth separation unless the complication of a torque load at output is added. The main use of accelerometers is rather specialised for conditions where tooth frequencies are high, above 1 kHz and torque is applied or where the lack of a free shaft end rules out encoders. They can however be fitted inside gearboxes whereas encoders are less likely to be oilproof.

The choice of encoders is controlled by factors such as the mechanical limitations on centre distance and the mass and speed limitations of large encoders. In practice, for gear noise work, accuracy is not a limitation. Choice between shaft mounting and coupling drive is not clear and depends on available space and on whether a robust shaft extension is available.

For the electronics, choice between the batching, velocity and high speed approaches is much more difficult. At metrology speeds, typically below 25 rpm, any of the systems can be used and will give satisfactory results and accuracies are comparable as it is the encoder accuracies which control the final result. At these speeds the choice will tend to depend on availability and cost of the equipment. The batching approach is probably the most straightforward if unskilled labour is doing routine production testing. However the equipment commercially available [9] is expensive as it is designed to handle a very wide range of test gears and to be "foolproof". The velocity approach is probably the cheapest option as it does not involve interpolation and only needs a standard data logging card in its simplest version but is slow. To increase speed sufficiently to work in real time a specialised counter card is needed.

At high rotation speeds it is not possible to use the batching approach and either the velocity or high speed approach must be used. The velocity approach requires fast computing ability and so tends to have to work off-line. The high speed approach has the ability to display the results in real time but initial zeroing is required, taking a few seconds if done manually or about 4 revs of the input if under computer control.

If the drive is not synchronous the double divide system does not like drift and the attendant complications so it is simpler to use the velocity approach or demodulation of the individual encoder signals. Demodulation using phase-lock loops works fast and effectively but is not easily or quickly altered if loop natural frequency has to be changed so although it is very suitable for test rigs which are always operating in a narow band of conditions it is less suitable for wide ranging conditions.

References

1.   Munro, R.G., 'A Review of the Theory and Measurement of Gear Transmission Error.' Int. Conference on Gear Noise and Vibration, I. Mech. E., April 1990, p 3.
2.   Heidenhain Ltd., 200 London Rd., Burgess Hill, Sussex, RH15 9RD, U.K. or 115 Commerce Drive, Schaumburg, IL 60173, U.S.A.
3.   Sweeney, P.J. and Randall, R.B., 'Gear transmission error measurement using phase demodulation.' Proc. Inst. Mech. Eng., Vol 210C, 1996, pp 201-213.
4.   Remond, D., 'Practical performances of high-speed measurement of gear transmission error or torsional vibrations with optical encoders.' Meas. Sci. Technol. 9 1998, I.O.P. pp 347-353
5.   Tuma J., 'Phase demodulation in angular vibration measurements.' International Carpathian Control Conference, Malenovice, Czech Republic. May 2002. (Dept. Control Systems and Instruments, VSB Tech Univ Ostrava, Ostrava, Czech Republic. jiri.tuma@vsb.cz)
6.   Smith, J.D., 'Gear Transmission Error Accuracy with Small Rotary Encoders.' Proc. Inst. Mech. Eng., vol. 201, No. C2, 1987, pp 133-135.
7.   Smith, J.D., 'A Modular System for Transmission Error Testing.', Proc. Inst. Mech. Eng. vol. 202, No. C6, 1988, p 439.
8.   Smith, J.D., 'Practical Rotary Encoder Accuracy Limits for Transmission Error Measurement.' Proc. Inst. Mech. Eng., 1991, 205 (C6), pp 431-436.
9.   Klingelnberg Ltd. PSKE 900. www.klingelnberg-oerlikon

8

Recording and Storage

8.1 Is recording required?

For much work, especially for initial investigations and development, there is little point in recording masses of data, whether T.E. or vibration. Displaying the information directly on an oscilloscope, preferably triggered to synchronise with 1/rev of pinion or wheel is very valuable and should never be omitted. It is especially useful when the problem occurs at particular points in the revolution. A typical example is the noise of a timing drive clatter on a diesel engine.

Even more important is the information from the raw signal to see whether noise or vibration is due to isolated impulses or to steady excitation. Steady vibration, typically at one-per-tooth frequency, is easily recorded by hand since the frequency is obvious and there is a single Figure for amplitude. A T.E. trace such as that sketched in Fig. 8.1 will give an immediate value for eccentricity and for the (expected) 1/tooth so no data logging is required, whereas a trace such as that in Fig. 8.2 needs recording for detailed analysis.

If a condition is transient (e.g., scuffing) or if there is a suspicion that a small regular defect is hidden underneath steady or irregular vibration, then it is essential to record for detailed subsequent analysis. It is not unknown for the signal-to-noise ratio to be -20 dB (or even lower) in a gearbox.

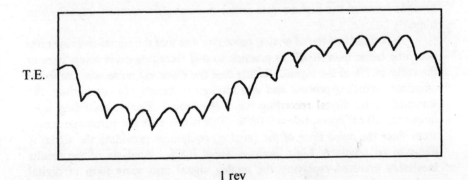

1 rev

Fig 8.1 Simple T.E. trace.

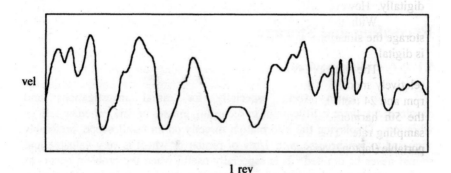

1 rev

Fig 8.2 Complicated vibration recording.

"Noise" in this context is used to describe any electrical or mechanical vibration which is not the vibration of interest.

8.2 Digital versus analog

Until 20 years ago analog (tape) recording completely dominated the field of data recording. Digital storage was expensive and restricted in size and sampling rate, so there was virtually no competition to 14, 16 or 32 track recording on magnetic tape. Information rates up to 300 kHz per track were possible, equivalent on a 14 track recorder to a total digital sample rate of well over 10 million samples per second. Total storage times were 700 seconds (even at the highest data rates) so equivalent memory capacity was huge.

A disadvantage of analog recording was that the signal-to-noise ratio was little better than 40 dB in practice so that recording noise levels were of the order of 1% of the signal. In this case the electrical noise was due to the magnetic recording process and was random in nature. In comparison, the standard 12 bit digital recording has a theoretical effective recording level more than 70 dB down, below 0.03%. This is not quite the advantage it may seem since the noise floor of the (analog) equipment providing the signal is likely to be relatively high, perhaps about 0.5%. Analysis of the results inevitably involved replaying the analog signal into some form of digital analysis equipment so that there was an extra transfer needed.

Current tape recorders are a hybrid since they typically record on video cassettes and can record multiple tracks at high rates but, like CD players, they record the information digitally. To replay, they convert the information back into analog form and it is then re-digitised in a computer for analysis. Signal-to-noise ratios are good since the information is stored digitally. However, such recorders are expensive and heavy.

With the advent of cheap active memory and very cheap digital storage the situation has now changed completely so that nearly all recording is digital.

The requirements for most gear noise and vibration work are relatively modest. Necessary recording frequencies are limited since 1450 rpm and 24 teeth is less than 600 Hz tooth frequency and we can record up to the 5th harmonic of this tooth frequency (giving 3 kHz) with a 10 kHz sampling rate. This leads us to record directly into a standard (cheap) PC or portable (laptop) computer.

8.3 Current PC limits

Given sufficient expenditure there are now few limits on what can be achieved digitally with a special purpose computer. However, prices rise very rapidly if we depart from what is standard and easily available so it is advisable to tailor testing to current standard PC performance.

A standard PC together with a basic 16-channel 12 bit data logging card can cost less than £1000 ($1500). It is not necessary to use an expensive card with output capabilities or sophisticated facilities. This will allow total sampling rates up to 200 kHz (kilo samples/sec) and the information can be poured (streamed) straight onto hard disc. The information is in the form of 12 bit samples so with direct storage each data point takes up 2 bytes of memory. A free memory capacity of 20 gigabyte on the hard disc allows 10,000 million samples to be stored and with 6 channels at 10 kHz (60 kHz in total) the recording time possible is 160,000 secs or 44 hours, far more time than is needed for a set of tests for noise investigation or development purposes.

If condition monitoring is being investigated then 44 hours is likely to be insufficient and techniques are needed to reduce the quantity of information to be stored.

Twelve bit resolution (1 part in 4096) is currently standard and is a good compromise. Eight or 10 bit resolution is not really sufficient when the signal contains a small vibration of interest, swamped by a large vibration that is irrelevant. Sixteen bit resolution is not needed since, with the fairly standard range of ± 5 V, each bit would be only 0.15 millivolts, well below the noise level. Resolution or discrimination, typically 2.4 mV for 12 bit recording, should not be confused with accuracy which is usually about 1% for vibration, equivalent to 100 mV for 10 V full scale. In general, absolute accuracy is not important because we are looking for changes or differences. Occasionally it may be worthwhile to consider double recording information, once with all the information present and then in parallel, cutting out irrelevant high or low frequency information with a filter and amplifying to give just the information of interest.

For data logging on site the same considerations apply, although the portable laptop computer and the necessary PCMCIA card are slightly more expensive, so the cost approaches £1500 ($2000) for up to 16 channels at 200 kHz total sampling rate.

It is tempting to consider streaming the test data straight onto CD instead of onto hard disc and there is then the advantage that if non-rewriteable discs are used there is a permanent very cheap archive. With a storage capacity of 650 MB or 300 M samples for less than £2 ($3) storage costs are negligible.

When T.E. is being recorded the requirements are for perhaps 4 revs at 1,000 samples per rev with 3 channels being recorded so each mesh check requires only 24 kB of storage. One CD can store the results for 20,000 gear checks.

8.4 Form of results

A question often asked is whether vibration information should be recorded, analysed or stored as acceleration, velocity or displacement, and there is sometimes frank disbelief that an acceleration signal, when integrated, provides a velocity signal.

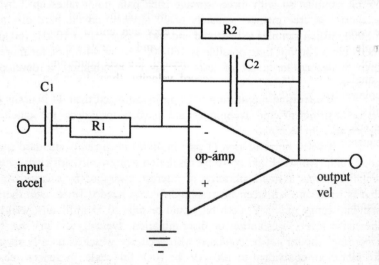

Fig 8.3 Circuit to integrate acceleration to velocity.

Almost exclusively, the original vibration measurement is now acceleration but it is easy to carry out one stage of integration to velocity, as in Fig. 8.3, with an operational amplifier.

The basic integration is the input resistor R_1 working with the feedback capacitor C_2 but an extra blocking capacitor is needed at input, and a parallel resistor R_2 in the feedback, to prevent drifting to saturation. The time constants (RC) for input and feedback should be kept larger than the value of $(1/\omega)$ for the lowest frequency to be measured. Typically the combination of an input R_1 of 100 kΩ and C_2 of 0.01 μF gives a time constant of integration of 1 millisecond so that if the input scaling is 1 V per m s^{-2} the output corresponds to 1 V per mm s^{-1}.

At input, an R_1 of 100 kΩ and C_1 of 1μF gives a low end rolloff frequency of 10 rad/s or 1.6 Hz and to match this with C_2 of 0.01 μF requires an R_2 of 10MΩ. If only audible noise matters, then the low-cut blocking frequency can be set fairly high at, say, 30 Hz, greatly reducing drift problems.

In theory a second stage of integration, identical to the first stage could be used to give displacement, but in practice this is rare. The double integration tends to give a rather unstable fluctuating signal which floats considerably since the slightest spurious components at low frequency in the original signal are greatly amplified by the double integration. Using chopper stabilised instrumentation amplifiers helps but does not completely solve the problem and may inject chopper frequency noise.

Integration can be carried out digitally on the signal but suffers from the same drift problems as the analog approach and a standard PC with simple software cannot stream data to disc and integrate simultaneously. If double integration to displacement is needed, the best compromise is usually to analog integrate to velocity, record velocity, then digitally integrate to displacement and then high-pass-filter to cut out spurious low frequency drifts. A convenient alternative is to record velocity and to frequency analyse the velocity signal then digitally divide each band amplitude by the angular frequency to get the frequency spectrum for the displacement.

Whether acceleration, velocity or displacement should be recorded depends on the engineering requirements. For noise purposes it is velocity that tends to be proportional to noise and it is also velocity that is most likely to remain roughly constant over a very broad range of frequencies. Hence, for noise investigations we usually record (and analyse) velocity using an analog integrator to avoid integrating digitally. This greatly reduces the danger of the signal of interest being too small, unlike using acceleration which is tiny at low frequencies or displacement which is miniscule at high frequencies.

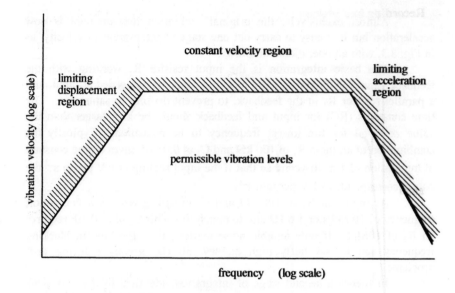

Fig 8.4 Typical test limit vibration specification.

In contrast, when positional accuracy matters for timing gears or printing, the low frequency components dominate the results and it is better to record displacement (as with T.E.).

For monitoring, the troublesome occurrences exist for very short time scales and acceleration is preferred, emphasising the higher frequency components. In extreme cases it can be worthwhile to consider recording "jerk," the differential of acceleration.

A typical "customer acceptance vibration specification" for a gearbox imposes a constant velocity limit (7.5 mm s^{-1} peak) over the central working part of the range, then goes to constant displacement limit (40 μm p-p) at low frequency and nearly constant acceleration limit (50 - 100 m s^{-2}) at high frequency (see Fig. 8.4, which is typical of the AGMA specification) [1,2].

This type of approach tends to assume that the problems exist at well separated frequencies so the separate frequency bands do not combine to generate high peak values. This is usually relevant for noise, but not when accuracy is involved, since a signal plus harmonics can give a peak value many times higher than a single component when pulses occur (see section 9.3). It is unfortunate that there is no easy method of substituting for a look at the original time trace on an oscilloscope. Humans are very good at detecting that something is different or "wrong" even though they may not be able to specify the problem exactly.

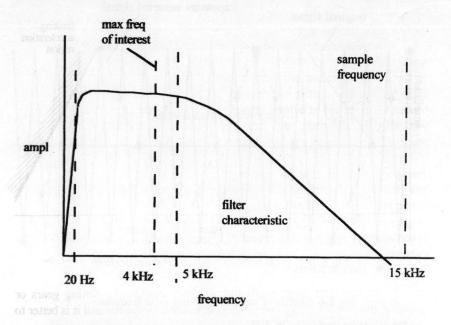

Fig 8.5 Typical frequency ranges for data recording and sampling.

8.5 Aliasing and filters

There is a very large amount of literature about electrical "noise" problems and about the problems of filtering, sampling and aliasing. Unfortunately not all that is written is necessarily correct when tackling a particular problem and high costs can be associated with sophisticated filters, which may be redundant.

The first essential is to decide on the frequency range of interest and a standard conventional solution is as indicated in Fig. 8.5. The (audible) frequencies of interest might be 30 Hz to 4 kHz, filters (band pass 4 or 6 pole) would be set at perhaps 20 Hz and 5 kHz, and sampling might be at 15 kHz (or technically 15 k samples/sec).

The sampling rate and filtering are interlinked. Sampling theory [3] says that we can detect a signal up to half the sampling frequency but the effect of "aliasing" is to allow false indications if there is high vibration above half sampling frequency. The effect is sketched in Fig. 8.6 and shows how a high frequency input at f_1, when sampled at f_s, can appear to be at a frequency of $(f_s - f_1)$. This means that vibration above $f_s/2$ needs to be filtered out.

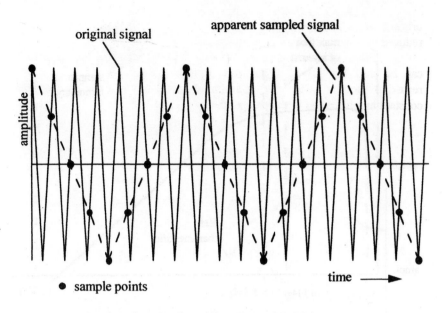

Fig 8.6 Sketch of sampling giving false frequency.

The effect is sometimes called a "picket fence" effect and is occasionally seen in very old films where car wheels appear to be rotating backwards. It is the same effect as using a stroboscopic flash to slow down or reverse a vibration or rotation.

The resulting frequency spectrum is "reflected" in the output spectrum as if there were a mirror at frequency $f_s/2$ (the "folding" or Nyquist frequency) and it means that a high signal at frequency 0.6 f_s will appear at a frequency 0.4 f_s, as in Fig. 8.7.

The mathematics of Fourier frequency analysis with sampled vibrations cannot detect the difference between those frequencies above $f_s/2$ and those below. When a fundamental frequency analysis is carried out, the result gives both the components above and below the folding frequency as conjugate pairs and we arbitrarily (and sometimes incorrectly) assume that it is solely the lower frequency that is there.

The job of the band pass filter is to make sure that all frequency components above $f_s/2$ are negligible so that they cannot influence the frequency range of interest. Filters are not perfect devices and if we take the standard (rather expensive) four pole filter it will have reduced amplitude by 2^4 at double its nominal or roll-off frequency. In the case quoted above with f_s at 15 kHz, a spurious signal at 10 kHz would be reduced to 6% of its value by a filter set at 5 kHz and would appear to be at a frequency of 15 - 10, i.e., 5 kHz. To appear within the frequency range of importance, < 4 kHz, the original vibration would have to be at 15 - 4, i.e., 11kHz, and would be reduced by a factor of $(2.2)^4$ (i.e., down to 4.3% of its original value).

Filters with a higher roll-off rate than the standard four pole filter can be used but they may be more expensive, more temperamental with regard to "ringing" when there is an impulse, and may give "ripples" of non-constant amplification in the passband.

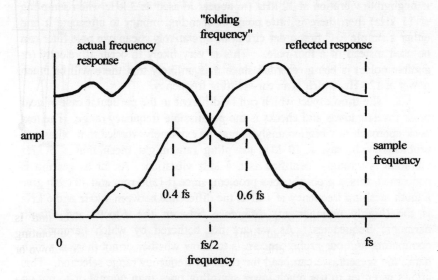

Fig. 8.7 Aliasing effect in sampled signal analysis.

For general testing the normal solution is to take the top frequency of interest f_g, set the high cut filter perhaps 25% above the top frequency, and set the sampling rate to $4 \times f_g$. The low cut filter is set slightly below the lowest frequency. This "standard" solution tends to be applied without much thought to all problems and is likely to result in a test setup that is unnecessarily expensive. The set of filters may easily cost more than the computer and data logging card and be an additional weight to carry and correspondingly increase equipment sales profits greatly.

The first casualty of actually using intelligence about the filter requirements is the need for a high performance (expensive) low-cut filter at the bottom end of the frequency range. A simple blocking capacitor will cut off DC and, especially if we record velocity, the time constants of the integrating circuit can be set to reduce the 1/rev components which, in any case, will be very small for both acceleration and velocity and will be ignored in the final assessment. This one change can halve the cost of filtering as well as increasing reliability. In one very large industrial monitoring installation, very expensive low frequency filters were used to cut out tidal effects, not only greatly increasing costs but removing a very useful permanent running check that the equipment was performing satisfactorily with regard to both timing and amplitude.

The second casualty can be the need for a relatively high performance (4 or 8 pole) filter at the top end of the frequency range. If there is negligible vibration at 12 kHz (to appear aliased as 3 kHz when sampling at 15 kHz) then there is little point in spending money to attenuate it and either a simple R-C first order circuit or a relatively cheap two pole filter can be used instead of a four-pole.   This is very likely to occur if velocity (or audible noise) is being recorded since it is unlikely that there will be much power at 12 kHz, which is an ear-splitting frequency.

The third aspect which can be different in the particular case of gear noise investigations and checks is the permissible frequency range. The text book approach to vibration analysis may be extremely worried that "aliasing" problems with, say, a 10 kHz sampling rate might mean that a 6 kHz vibration is wrongly identified as a 4 kHz vibration. As far as gearing is concerned, this is probably not a problem, since if 1500 rpm and 40 teeth give a tooth meshing frequency of 1 kHz, the difference between 4 kHz and 6 kHz is the (highly unimportant) difference between the 4/tooth and 6/tooth harmonic frequencies.   As we are not bothered by which harmonic is dominating and our prime concern is to know whether or not there is a high harmonic present, we can bend the rules on frequency range selection. This allows us either to use much lower sampling rates than normal or to put up the detection range relative to a "standard" sampling rate. In one particular gear monitoring problem the sampling rate was set at a predetermined 10 kHz so use of the standard approach would have limited the high cut filter to about 3 kHz. The high cut filter was in fact set to 7 kHz so that instead of the information being limited to 3rd harmonic of tooth frequency there was a very useful (if, technically, possibly incorrect) information recording and detecting up to 7th harmonic. When replaying it, the 7th harmonic would show as the 3rd, and the 6th as the 4th, but this was not important as the objective was purely to detect trouble, not to identify it accurately.

When a compact (cheap) system is desirable filter chips are available typically giving a 5th order Butterworth characteristic and two such chips can be cascaded to give high rolloff rates cheaply. They need to be driven by a TTL oscillator such as an 8038 at 100 times their required rolloff frequency. There is also a limitation that the maximum input voltage is limited to about 4 V when the rails are at 5 V. This requires that an input signal is reduced to below 4 V, filtered then re-amplified to return to the original size.

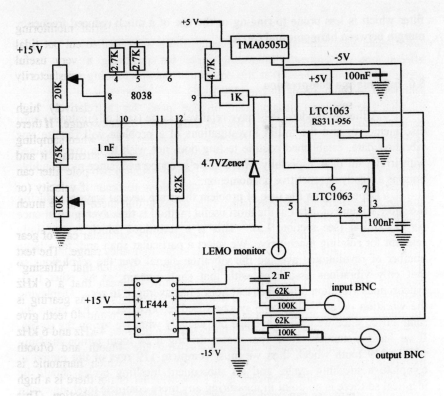

Fig 8.8 Circuit for double 5th order low pass filter.

Such a circuit, as shown in Fig. 8.8, is not very accurate for its rolloff frequency and is restricted in its performance but is sufficient for portable T.E. measurement purposes and can easily be fitted onto a standard board to give a very compact unit for travelling.

There is a trade-off between filter performance and sampling rates which can occasionally be of help in T.E. testing where there is a large but irrelevant additional signal present. With the high speed double-divide system, the carrier frequency tends to be fixed by the requirement to give enough full scale to accommodate eccentricities in less accurate gears. The 5th harmonic of tooth frequency will also be fixed and there may then be a low (< 3) frequency margin between the harmonic (at < 1 μm p-p) and the carrier (at 400 μm p-p). To prevent aliasing when sampling at normal rates requires attenuation greater than 60 dB but if the sampling rate is increased to above twice the carrier fundamental frequency the carrier will appear in the final frequency analysis but will appear at its correct frequency and so can be ignored. This allows the use of a lower performance and hence more stable filter which is less prone to ringing or the use of a much reduced frequency margin between harmonic and carrier.

8.6 Information compression

Although modern PCs have relatively large (tens of gigabytes) hard disc memories and the initial investigations of a problem will require raw vibration data, established routine testing does not wish to be overwhelmed with irrelevant data, especially where noise is concerned, since most audible noise is a steady or repetitive phenomenon.

Depending on the type of problem there are several ways of reducing the sheer volume of data but the most useful method is time averaging at once per revolution (see section 9.5). This is a technique which is especially relevant for rotating machinery. We select a particular shaft and, for a large number of revolutions, average the vibration signal over the revolutions so that only vibrations associated with that shaft remain, as all other non-synchronous vibration (and electrical noise) has averaged to zero. Displaying the vibration on an oscilloscope synchronised to once per rev has much the same effect since we tend to average out visually what we see on the screen. If we have a standard 1500 rpm motor driving a 24 tooth pinion meshing with a 119 tooth wheel, then we must complete 119 revs of the pinion to complete a meshing cycle, and all subsequent meshing cycles should be identical so there is no point in measuring any more complete meshing cycles since the same information should appear again and again. This will take 4.76 seconds for the cycle, and with tooth frequency 600 Hz and a requirement to measure up to 7th harmonic we would sample at perhaps 16 kHz. A complete meshing cycle is then 76,160 data points for each of the channels recorded.

At the operating speed, a single revolution of the pinion (40 milliseconds) corresponds to 640 data samples and a single revolution of the wheel corresponds to 3173 samples. Since all the information relevant to the complete meshing cycle can be stored as one averaged revolution of the pinion and one of the wheel, we only need to store 3813 items of information instead of 76,160. Any other method of storing all the information relevant to a complete meshing cycle either requires much more storage or is much less accurate. It is usually assumed that storing vibration information as a frequency analysis is much more compact than storing the original raw information, but this is not correct for the semi-repetitive information we get with machinery. It is only correct if debatable assumptions are made about a stationary noise spectrum [3].

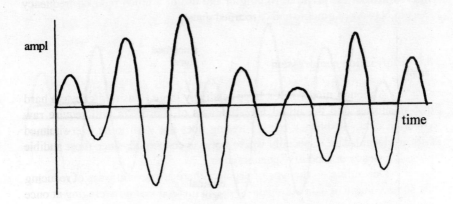

Fig. 8.9 Rectification of vibration signal.

Another possibility for information compression arises when we already know that the signal consists of a limited number (usually just one) of (known) frequencies. We can then use "enveloping" techniques which give us the overall amplitude of vibration without bothering with the detail of each individual cycle. The sampling rates needed for the envelope are much lower than for the original vibration.

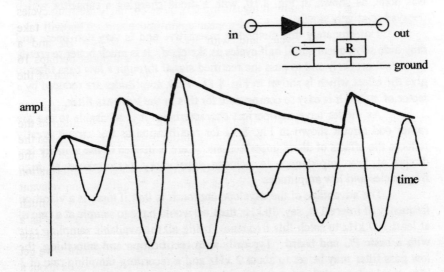

Fig. 8.10 Former method of enveloping vibration signal.

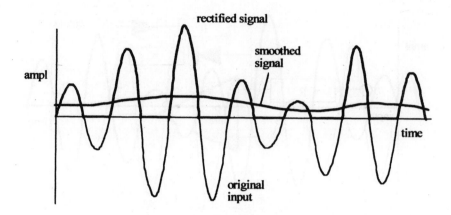

Fig. 8.11 Preferable method of enveloping.

This type of information may be relevant for looking at 1/tooth frequency and its modulation due to varying misalignment or torque effects or looking at high frequencies when damage monitoring as the ringing of an accelerometer may be triggered by metal to metal contact (see Chapter 15)

Fig. 8.9 shows how the vibration signal, at a single frequency, symmetrical about zero, is rectified ready for "enveloping." Originally this was done, as shown in Fig. 8.10, with a diode charging a capacitor which decayed relatively slowly.

Unfortunately this method is insensitive and is very non-linear and may hide subsequent small half cycles as sketched. It is much better to rectify the signal properly and to pass the rectified signal through a low pass filter to give the effect which is shown in Fig. 8.11. Peak amplitudes are reduced by a factor of π but it is easy to compensate for this in the low pass filter.

As diodes have non-perfect characteristics it is advisable to use the rather odd circuit shown in Fig. 8.12 for rectification as this circuit greatly reduces the effects of diode imperfections. Care is needed to use suitably fast diodes at low impedances to achieve satisfactory performance at high frequencies and low amplitudes.

The advantage of the envelope approach is that if there is a vibration frequency of interest at, say, 30 kHz then we would have to sample at a rate of at least 100 kHz to catch this frequency, using all the available sampling rate with a basic PC and board. Typically with rectification and smoothing, the low pass filter may be set to about 2 kHz and a recording sampling rate of 3 kHz would be satisfactory, despite the normal sampling rules quoted in textbooks.

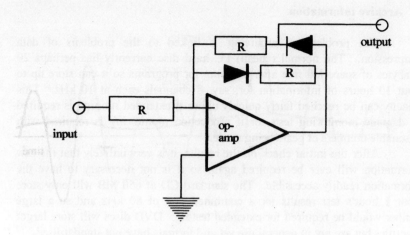

Fig 8.12 Circuit for accurate rectification of small signal.

The standard sampling rules do not seem to apply for problems such as this where the main requirement is to have the area under the envelope roughly right. Practical testing with an artificially generated signal with bursts of perhaps six cycles of vibration and testing by varying filter frequency will give a very clear visible check on what frequencies of rolloff and sampling are satisfactory. Such a test signal can be obtained by (analog or digital) multiplying a single sided square wave by the carrier (30 kHz) to give a test signal similar to the expected signal. The resulting reduction in sampling rate and, hence, data storage due to enveloping is typically at least 30:1.

Another possible method of reducing information storage is to take advantage of the known form of the structure of a frequency analysis of a repetitive waveform such as that from a gear set. We know that as the waveform is repetitive there can only be frequencies at exact multiples of once per revolution and that for most gears with whine noise problems it is only the 1/tooth frequencies and harmonics that are relevant. There is then no point in recording amplitudes of all frequencies from the Fourier analysis as there are only perhaps five frequencies that are relevant for a typical back axle whine.

In section 9.3 the possibility of amalgamating several lines from a frequency analysis of a T.E. record is mentioned as an aid to having a clearer assessment of the total power in the region of a tooth frequency or harmonic. This also reduces the amount of information stored (by a factor of 10) if it is being stored in the form of a frequency table.

8.7 Archive information

The problem of archiving is linked to the problems of data compression. The normal (cheap) PC hard disc currently has perhaps 20 gigabytes of space left free after allowing for programs so it can store up to about 30 hours of information for, say, 8 channels each at 10 kHz. This capacity can be reached fairly quickly either if extended running is required for damage monitoring tests, or if production monitoring is required with reasonable numbers of gears being made.

After the initial check on the results, it is very unlikely that the raw information will ever be required again so it is not necessary to have the information readily accessible. The standard CD at 650 MB will only store about 1 hour's test results for a combined rate of 80 kHz and so a large number would be required for extended testing. DVD discs will store larger quantities but are not in general use yet and formats have not standardised.

A suitable compromise for vibration work or T.E. tests is to store selected small files such as the most interesting mesh cycle averaged files (as in section 8.6). These averaged files contain typically less than 4 k points and so are only 8 kB, allowing noise test results from thousands of tests to be stored on a CD Rom, easily accessible for quick checks.

The problem is then whether or not to bother storing the original raw data which takes up perhaps 10 MB per test, just in case there are strange intermittent irregularities in the results which do not necessarily appear in the averaged traces. Caution suggests that, like taking out insurance, the information should be archived just to ensure that it will never, ever, be needed. Most PC systems have some form of backup and typically 650 MB backup costs less than £2. Since it is wise to have such a system to backup the 400 MB of software on a computer, it is also wise to use it for archiving test data. It does not matter whether the CD writer is fitted internally or, as is more likely with a laptop, externally via USB. It is worthwhile using non-rewriteable CDs to remove the temptation to reuse discs as well as this being more economical.

There are much more elaborate, very high capacity backup systems usually based on tape drives but it is not worthwhile installing a system solely for archiving test records.

Linked to the problem of generating archives in the first place is the almost impossible problem of deciding when information should be scrapped. There is little point in storing information on gears which have already worn out but it is extremely difficult to take a decision on the time scale for killing off old records. This is one problem to which there is no satisfactory solution but the more compact the storage the longer can the decision be delayed.

References

1.   American Gear Manufacturers' Association. AGMA Standard 6000-A88.
2.   Dudley, D.W., Dudley's Gear Handbook, Ch 13 Gear vibration, McGraw-Hill, New York, 1992.
3.   Newland, D.E.N., Random vibrations, spectral and wavelet analysis. Longman, Harlow, UK and Wiley, New York, 1993.

9

Analysis Techniques

9.1 Types of noise and irritation

One of the most difficult problems in gear noise investigations is that the final "detector" and arbiter (on whether or not a noise is irritating) is an extremely non-linear, rather temperamental, and extremely variable human being, with office politics and economics playing a major role. It is quite possible for three people to listen to a gear drive and to object to it for three completely different reasons. No amount of technical measurement will determine which aspect of a gear drive noise will irritate a particular customer, so it is most important to identify the problem correctly at the start by questioning the customer thoroughly and by possibly playing tapes of different types of gear noise to the customer for comparisons. A PC with an output card to a loudspeaker can be useful for this.

There are, roughly speaking, four types of irritation:

(a) A steady tone. This is relatively musical and, because there are few harmonics, sounds a bit like an oboe. It is often encountered as a "back axle whine" on rear wheel drive cars and is typically in the 500 - 1000 Hz range (900 rpm and 40 teeth). A higher harmonic content moves the character towards a stringed instrument sound.

(b) A modulated tone. Here the customer is not objecting to the steady component at perhaps 400 Hz but to the fact that it is modulated (or wowing) at a much lower frequency. It is not uncommon to have a customer complaining that he is hearing a noise at 2 or 3 cycles a second. This is impossible. What is heard is the basic 400 Hz once-per-tooth noise being modulated in amplitude (or phase) at 2 or 3Hz.

(c) 1/rev impulses. This is the type of noise generated by a defect such as a nick or burr giving an impulse at 1/rev and is usually most noticeable at low speeds. The sound is a fast ticking sound and has very little power associated with it so it will not usually show up in a frequency analysis. However, like a triangle in an orchestra, it can easily be picked out by the peculiar non-linear abilities of the human ear.

(d) Grumbling or graunching. This is the "classic" gearbox noise, usually associated with low speed and heavily loaded drives. It is the typical "bottom gear" noise in a car. It tends to be associated with pitch errors and is essentially at all harmonics of once per revolution of both wheel and pinion. Frequency analysis is of little help since all frequencies (or all multiples of a couple of very low frequencies) are present.

Which of these types of noise causes the irritation depends, to a large extent, on what the listener is expecting. One engineer will often expect (a), (b), and (d) and ignore them but will be highly irritated by (c), whereas another might reject due to (b). One car driver might be irritated by (a) and ignore (d), while another would react the opposite way. Occasionally, as with a car, it is not the noise itself which irritates but the fact that the noise has changed from a familiar, accepted "normal" noise.

There is interaction in human response between the various sounds and sometimes it is possible to use the deliberate addition of pitch errors in a drive to break up the sound pattern. This technique is sometimes used in chain drives if the customer is irritated by a steady whine.

9.2 Problem identification

From what has been said in section 9.1 the accurate specification of the problem is not always easy. Occasionally it is a simple pure tone that is heard and, if a quick check with a sound meter straight into a frequency analyser or oscilloscope (see section 6.2) confirms that the frequency is once-per-tooth, diagnosis is easy.

Checking the character of the sound is a great help and if the sound is complex, some form of artificially generated range of sounds can help identify the type of noise. This can be done using predominantly analog equipment but it needs quite a complicated setup so is more cheaply tackled by generating a series of repetitive time sequences with and without the various errors in a standard PC. The resulting time series for each revolution is then fed via an output card into an audio amplifier and loud speaker or can be played out on a sound card. The problem with standard soundcards is that varying the frequency is not easy. Reasonable resolution is obtained if each tooth interval is, say, 30 samples long and 25 teeth need 750 sample points per revolution.

The various types of error can be generated as (Fig. 9.1):

(a) 1/tooth errors.   amplitude times mod (sin $\pi x/30$) gives the typical half sine wave of 1/tooth (for $x = 1:750$ as the position round the revolution).

(b) Pitch errors. These can be put in as positive and negative at arbitrary positions of x. The classic dropped tooth can be modelled as h $x/750$ where h is the drop size. It is helpful to be able to either add or subtract a given pitch error because the audible effects are not necessarily the same.

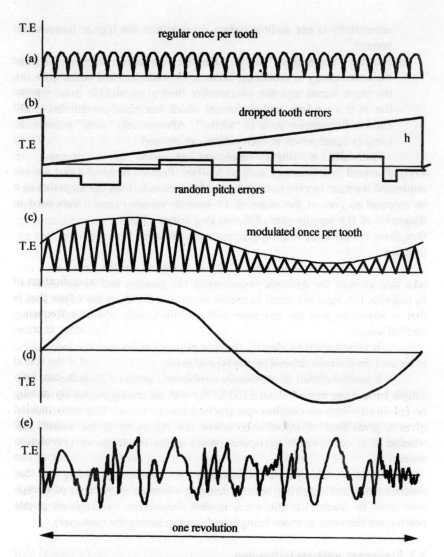

Fig 9.1 Models of various types of noise generated by gear drives.

    (c) Modulation. Multiplying the sequence of 1/tooth errors by $(1 + \sin (2\pi x/N))$ allows modulation at 1/rev (N = 750) or wheel frequency (N = 1300) or 2/rev (N = 375) for a diesel or at any other possible torque variation frequency.

    (d) Eccentricity. This can be modelled as $e \sin (\pi x/375)$ and added in but will not alter the sound. It is, however, useful for demonstrating that eccentricity is not audible unless it modulates the higher frequencies present.

(e) Random "white noise" can be added for comparison purposes. Again the terminology is muddling because we add electrical white noise to the input signal and the loudspeaker then gives audible noise which has in it a random content (noise) which has equal amplitudes at all audible frequencies so it is "white." Alternatively "pink" noise with roughly equal power in each octave can be used.

Generally a single revolution sequence in a program is straightforward in a language such as Matlab. Perhaps 60 revolutions can be sequenced together to give runs of the order of seconds, then the sequence can be repeated to give of the order of 10 seconds running time. Varying the frequency of the sample rate of the analog output channel on the computer then gives the effect of varying gearbox speed as when running a gearbox up to speed.

Using the original typical T.E. as the input for the noise does not take into account the dynamic responses of the gearbox and its installation. In practice, this does not seem to matter since it is the character of the sound that is important and the customer will usually readily identify the "same sort" of sound.

It is important to identify the type of problem because the techniques to be used for analysis depend on the type of error.

Equally helpful, as previously mentioned (section 6.2), is the use of a simple basic noise meter (about £100/$150) with an analog output which can be fed directly into an oscilloscope synchronised to 1/rev. This immediately gives a great deal of information about the regularity of the sound and whether it is occurring at particular points in the revolution or is a steady sound.

If the microphone information is confusing, going to an accelerometer and checking bearing housing vibration is the next move but care must be taken that the main trouble frequencies investigated at the bearing are the same as those being heard (and irritating the customer).

9.3 Frequency analysis techniques

Fourier ideas start with the observation that any regular waveform can be built up with selected harmonics with correct phasing. Fig. 9.2 shows how the first four harmonics (all sine waves) added start to approximate to a saw tooth wave. It is important to get the correct phasing of the harmonics relative to the fundamental or you get a completely different character of waveform.

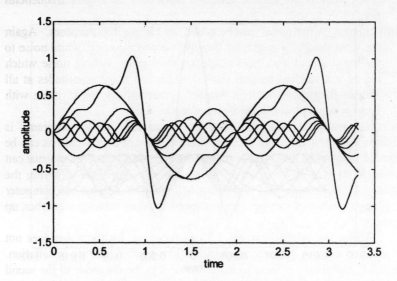

Fig 9.2 Build up of saw-tooth waveform with first four harmonics.

The technique which dominates most (digital) analysis currently is Fourier analysis, usually called fast Fourier transform (FFT) [1] because it is technically a computationally more efficient number crunching process than the classical multiplication technique. The details of the algorithm are irrelevant but it is worth noting that routines prefer to have an exact binary series number of data points; 1024 was popular but 8192 is now often used for irregular or non-repeating vibration.

However, if a signal has been averaged to once per revolution then it is the number of data points per revolution that must be used to get a correct answer and the sequence should not be "padded" with extra zeros.

This basic idea can be extended to a single occurrence such as a pulse. A pulse can be considered as one of a repetitive string with a very long wavelength so that the fundamental frequency approaches zero and "harmonics" then occur at all finite frequencies. Alternatively, a pulse occurs if a large number of waves of equal, but very small, amplitude happen to all have zero phase at a single point. At that point they will reinforce, giving a pulse, but at all other places will randomly add to (nearly) cancel out to zero. Fig. 9.3 indicates how the components build up.

If, however, the components do not all have zero phase at a single point in time the end result is a small amplitude random "white noise" vibration.

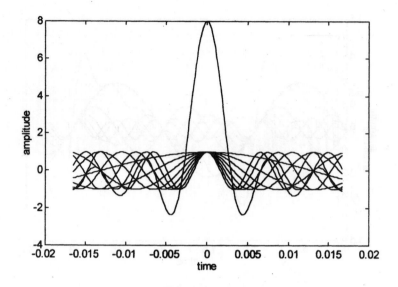

Fig 9.3 Seven components coinciding to give a pulse.

The reverse process involves using a sine wave as a detector by multiplying the signal under test by a sine wave of frequency ω (and unit amplitude) and averaging (or smoothing) the resulting signal.

Any component not at ω will average to zero over a long period since the product is negative as much as it is positive (Fig. 9.4), but if there is a component $A \sin\omega t$ hidden in the signal, then the output is $A \sin^2\omega t$, which averages to a value $A/2$. Initially the two signals in Fig. 9.4 were in phase so they gave a positive product, but then they became out of phase and gave a negative with cancellation over a long period.

Testing at all frequencies and with both sin and cosine detects all possible components. This classical approach involved testing over a longish time scale (with limits of integration $-\infty$ to $+\infty$) and returned an amplitude of a particular harmonic.

Current digital techniques work to a finite time scale (or to be precise, a finite number of samples) so they give a slightly different form of result. A finite number of points (formerly 1024) leads to the calculation of the total energy within a narrow frequency band whose width is determined by the number of sample points or the time scale of the test. As with all frequency analysis, in theory at least, the longer we sit and test, the more accurate the result and the narrower the measurement band possible. This is because the longer time scale allows the signals to change phase if they are not exactly the same frequency.

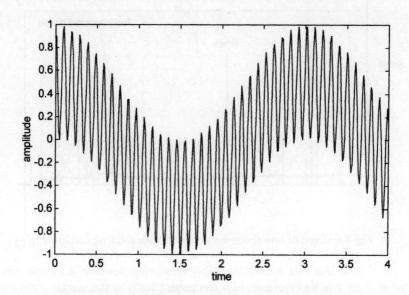

Fig 9.4 Result of multiplying two slightly different frequencies.

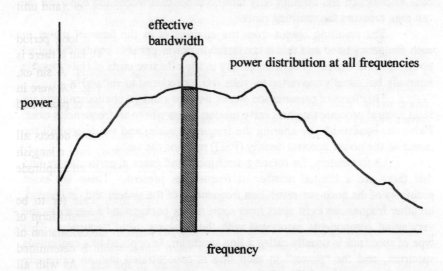

Fig 9.5 Frequency analysis with finite bandwidth.

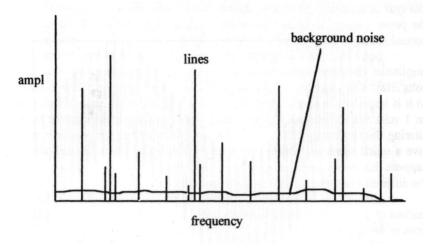

Fig 9.6 Type of line spectrum obtained with rotating machinery.

The idea that we are inevitably measuring power in a narrow band rather than a finite amplitude of a component leads to the mental picture in Fig. 9.5. Here we have many components at a range of frequencies and the effect of the analysis techniques is to model an almost perfect narrow band pass filter which lets through only those components within the band and we can then measure the resulting power.

The resulting output from the analysis is in the form of power in each frequency band and this is converted to power per unit bandwidth called power spectral density (PSD), originally in the effective units of (bits^2/sample interval) but usually converted to volts^2/Hz or reduced to volts/$\sqrt{\text{Hz}}$.

This form of presentation works well for random phenomena and for most natural processes such as wave motion at sea where all frequencies exist. Halve the bandwidth (by altering the frequency scale) and we detect half the power so the power spectral density (PSD) remains the same.

Unfortunately, for rotating machinery and gears in particular we find that there are a limited number of frequencies present. These are exact multiples of the once-per-revolution frequencies of the system and, in general, no other frequencies exist apart from some minor background noise and some very small components associated with the meshing cycle frequency. This type of spectrum is usually called a line spectrum, as opposed to a continuous spectrum, and the "power" in each line is concentrated into an extremely narrow frequency band (Fig. 9.6). A line will be at 29 times per revolution and at 29.1/rev there is technically no power though there will generally be power at 28 and 30/rev due to modulation of the 29/rev at once-per-rev. For this type of spectrum if we halve the bandwidth the PSD. will double since all the power resides in an extremely narrow line, well within the width of a normal band.

Some commercial equipment expects the user to be measuring line amplitudes (in volts) but most equipment expects to be measuring PSD. (in volts^2/Hz). Unfortunately, it is customary with both to give amplitudes in dB so it is important to check whether a reading is 25 dB down on 1 V (line) or on 1 volts^2/Hz (continuous). If, as usual, handbooks are uninformative, then altering the timescale with a single frequency input from an oscillator will give a quick check on which type of readout is being given. It sometimes happens that those manufacturing and selling the equipment are not aware of the difference between the two types of spectrum.

Conversion between the two types of readout is not difficult. Take a readout of -20 dB on 1 volts^2/Hz with a total bandwidth of 10,000 Hz and 400 lines in the spectrum. Each "line" is 25 Hz wide and the power is 0.01 V^2/Hz so the total power in that spectrum band is 0.25 V^2, which corresponds to a line amplitude of 0.5 V. In contrast, if the reading was -20 dB on amplitude the voltage would be 0.1 V and the PSD would be 0.01 V^2/25 Hz, i.e., 0.0004 V^2/Hz or 0.02 V/$\sqrt{}$Hz, which is -34 dB. The only time the two readings would agree would be if the bandwidth were 1 Hz. In practice the units may be given in g acceleration, mm/s velocity or μm displacement instead of volts but the conversion principle is the same.

Previous analog equipment for frequency analysis worked on the completely different principle of having a variable frequency tuned resonant filter which scanned slowly up through the range. This method is slow, expensive and not very discriminating and requires long vibration traces for analysis. It also has the disadvantage that tuned filter circuits do not respond rapidly to changes in vibration level. There is a digital convolution equivalent which can be used as a band pass filter (when modulation patterns are of interest) to extract a neighbouring group of frequencies, as occasionally happens with epicyclic gears, but it is rare for this to be required.

When a frequency analysis is carried out on a vibration or T.E. the band width of the resulting display is controlled by the testing time. Testing for 1 sec would give a bandwidth of 1 Hz for the output graph whereas a test for 0.1 sec gives 10 Hz bandwidth. This bandwidth may be unfortunate if it is too fine so that there are several lines associated with a particular frequency such as 1/tooth. The answer may be correct but it makes comparisons between different gears difficult or may give deceptive answers if the tooth frequency of interest happens to lie on the borderline between two bands as half the power will appear in each band.

148
Chapter 9

One possibility is to reduce the test time since halving the length of record will double the band width but this may mean that the test is for too short a time to give an average value over a whole revolution or longer.

A preferable alternative is to carry out the frequency analysis with the original (long) record then take the resulting Fourier analysis and add bands in groups. If the original record was for 10 s the bandwith would be 0.1 Hz and adding 10 bands would widen the bandwidth to 1 Hz. The addition is an addition of power in the bands so the modulus of the result in a given band must be squared, the band powers added, then the root taken of the sums. This is simply achieved in Matlab by a short subroutine such as

```
rrf=4*(fft(RST1))/chpts ;          % original record RST1 p-p values
trrf = abs(rrf(2:1001)) ;          % knocks out DC and gives modulus
pow = trrf.*trrf;                  % squares each line
frth = sum(reshape(pow,10,100));   % adds 10 lines to give 1 Hz band
sqfr = sqrt(frth);                 % gives p-p values for 10 lines
```

9.4 Window effects and bandwidth

One side effect of finite length digital records being used with frequency analysis is that the sudden changes at the ends cause trouble.

In Fig. 9.7, with a finite window length L, frequency analysis of curve A will give an exact twice per L sine component and no others, and curve B will give an exact twice per L cosine component and no others. Curve C gives trouble because the actual frequency, 1.2 times per L cannot exist in the mathematics, which can only generate integer multiples of frequency 1/L. The result of a frequency analysis on this wave is that the answer contains D.C. and components of all possible harmonics of 1/L. The result obtained is exactly the same as that obtained by analysing the repetitive signal shown in Fig. 9.8.

To overcome this problem in the general case of an arbitrary length record taken at random from a vibration trace, it is necessary to multiply the original vibration wave amplitudes by a "window" which gives a gradual run in and run out at the ends (Fig. 9.9).

This eliminates the sudden changes at the ends and greatly reduces most of the spurious harmonics generated as a consequence. There are various window shapes used, with the Hanning window being the most common. The various standard windows and their characteristics are described by Randall [2]. The side effect of using a window is that the effective length of the sample is rather shorter than the total length so the total power within the window is reduced and correction is made for this within the standard programs.

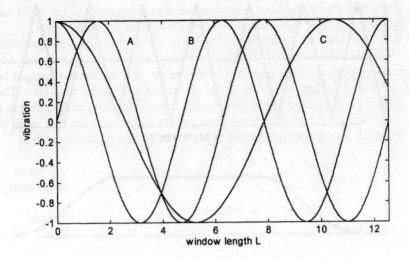

Fig 9.7 Finite length records showing end effects.

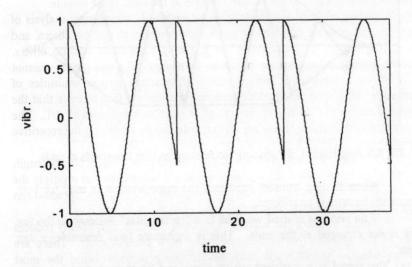

Fig 9.8 Equivalent continuous record for short sample.

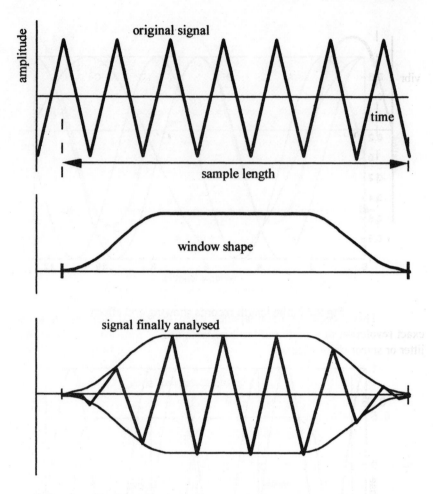

Fig 9.9 Application of "window" to remove sudden transients at ends.

When finding transfer functions, the same window is used for both signals so the ratio is unaffected.

If no window is used we refer to a "rectangular" window so the test data is not changed at the ends. This is legitimate (and desirable) either when:

(a) The signal is a transient which starts and finishes at zero (as when impulse testing a structure) or

(b) The signal contains only exact harmonics and so each harmonic component starts and stops at the same height with the same slope and the repeated signal (Fig. 9.10) would be smooth and continuous.

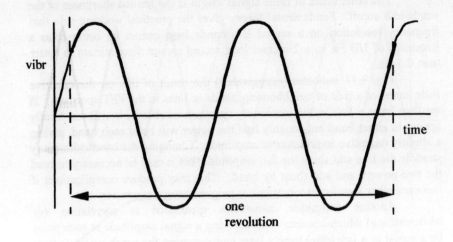

vibr

time

one
revolution

Fig 9.10 A signal that is correctly repetitive.

This second condition applies when the signal corresponds to an exact revolution of a shaft and has been obtained by time-averaging (without jitter or smearing problems).

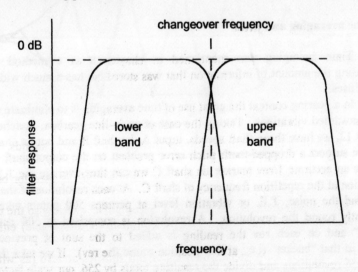

changeover frequency

0 dB

lower
band

upper
band

filter response

frequency

Fig 9.11 Problem of spectrum line at borderline between two filter bands.

The other effect of finite signal length is the limited sharpness of the bandwidth cutoff. Fundamental theory gives the practical working rule that frequency resolution on a record B seconds long cannot be better than a frequency of 1/B Hz so a 2 second long record cannot discriminate to better than 0.5 Hz.

Fig. 9.11 indicates (exaggerated) the result of this on the effective filter shape of a pair of neighbouring bands or lines in the FFT spectrum. If we then have a line in the machinery spectrum as shown, it will not totally appear in either band but roughly half the power will be in each band, giving a slightly deceptive impression of amplitude. Changing the bandwidths may straddle the line and show the full amplitude, but it may be necessary to read the two powers and add them by hand. This may produce complications if there are other significant vibrations at neighbouring frequencies.

Another technique sometimes mentioned is correlation (or whitewashing) which consists of multiplying a signal amplitude at each point by a signal at a (variable) time τ later and summing the result which is then plotted against τ. This technique is cumbersome but determines whether there is something "interesting" happening with a delay time τ. In the case of any rotating machinery we already know that "interesting" things happen 1 rev later so this technique is of little help and instead we use time averaging, which is much more economical of computing effort and more powerful as well as being faster.

9.5 Time averaging and jitter

Time averaging was mentioned in chapter 8 as a method of compressing the amount of information that was stored but has a much wider range of uses.

In a gearing context the great use of time averaging is to eliminate or reduce unwanted vibrations. Taking the case of an in-line gearbox, sketched in Fig. 9.12, we have three shaft speeds, input A, layshaft B and output shaft C. If we suspect a dropped-tooth pitch error problem on the output shaft C and have an accurate 1/rev marker on shaft C we can time-average the T.E. or vibration at the repetition frequency of shaft C. At each revolution of shaft C we read the noise, T.E. or vibration level at perhaps 500 points taken consistently round the revolution. A revolution is comprised of 500 data "buckets" and on each rev the reading is added to the sum of previous readings in that "bucket" (i.e., at that position round the rev). If we take the sum of 256 revolutions and divide the resulting totals by 256, our scale factor is unaltered and we have obtained an average vibration.

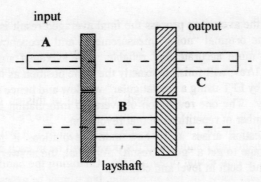

input

A

output

C

B

layshaft

Fig 9.12 Sketch of in-line gear drive with three shafts.

All vibration related to the output shaft, such as an output gear pitch error, will repeat in exactly the same place round the revolution so it will remain unaltered in size. All other non-synchronous, intermittent, random, or irregular vibration will behave like random vibration and average to zero.

Even powerful vibration such as engine inertia and firing effects from a 4-cylinder engine will be non-synchronous for the ouput shaft (though synchronous for the input shaft) and will be spread out round the revolution leaving mainly those vibrations associated with the output gear (and prop shaft and hypoid pinion if fitted).

A very narrow firing pulse consistently at one point on the input shaft will appear at each tooth interval on the averaged layshaft trace reduced in amplitude by a factor equal to the number of teeth on the layshaft. Fig. 9.13 shows the effects of a consistent narrow firing pulse of height H if it is on the "averaged" input shaft and if it is on a neighbouring shaft, in this case, the layshaft.

Additional to the benefit of extracting the information associated with a particular shaft rotation, averaging increases the accuracy of the readings and improves the resolution. If the original full scale (10 volts) is represented by 12 bits, then, after averaging, the total can in theory be up to 12 bits x 256 which is 20 bits size so after averaging we can have a 20 bit range. This seems to be impossible since, if full scale is 10 volts then originally 1 bit is 2.4 mV, and it does not seem possible to achieve a resolution better than 0.01 mV. In practice this can and does happen although only if there is extra random vibration present. For some accurate measurements a "dither" vibration [3] is deliberately added to increase accuracy of the averaged signal and to give resolution to the equivalent of better than 1 bit on the original measurement. In the case of a gearbox we have plenty of extra non-synchronised vibration around so we do not have to bother adding in the dither.

Due to the averaging process the final averaged result is much more accurate than the original "noisy" measurements and frequency analysis is correspondingly more accurate and reliable. An averaged signal, exactly synchronised to 1/rev, will finish at exactly the same position as it started and can be analysed by FFT using a "rectangular" window and hence using all the information fully. The one revolution of averaged information is equivalent to an infinite number of repetitions of that revolution.

The question arises as to how many revolutions is necessary or desirable to average to get a "good" result. As usual, the answer depends on the "noise" around, both in level and character.

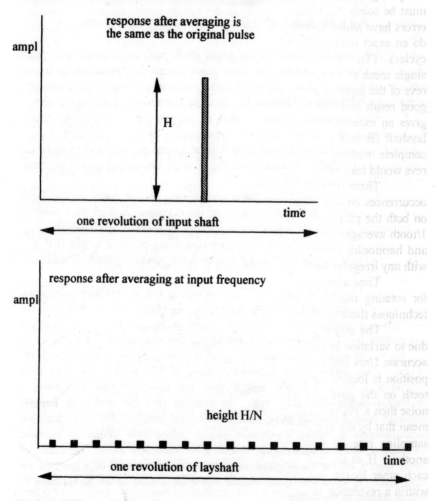

Fig 9.13 Effects of averaging at once-per-rev of input shaft and of layshaft.

Whether audible, mechanical or electrical, random noise which is comparable in size to the signal of interest is likely to be reduced to negligible importance (1%) if we average 128 cycles. If, however, the noise is 20 dB greater than the signal we may have to go to 1024 averages but this, fortunately, is rare.

When the "noise" is due to pitch errors on a mating gear, then the requirement is slightly different. By definition, the sum of all adjacent pitch errors on a gear must be zero since, otherwise, we do not finish a revolution where we started. If, then, one selected tooth on a pinion mates once with every single tooth on a wheel (with N_w teeth) then the sum of all the errors must be solely N_w times the error on the pinion tooth, since all wheel pitch errors have added to zero. To get the best averaging on the pinion it should do an exact multiple of N_w revs (i.e., an integral number of complete mesh cycles). (This assumes that, as usual, there is a hunting tooth.) So, with a single mesh of 19 teeth (input) to 30 teeth (output) we need multiples of 30 revs of the input to give complete meshing cycles and 120 revs would give a good result and reduce random noise effectively. This meshing cycle idea gives an excessive requirement if there are two meshes, as happens with a layshaft (B in Fig. 9.12) with 19:29 at input and 23:31 at output. For a complete meshing cycle the layshaft would have to do 19 x 31 revs and 589 revs would take rather a long time and require some 300,000 data points.

There is an exception to the basic idea that time averaging separates occurrences on two meshing shafts. Regular 1/tooth and harmonics appears on both the pinion and wheel averaged traces since the steady component of 1/tooth averages up consistently. In both averaged traces the regular 1/tooth and harmonics components associated with both shafts will appear together with any irregular tooth components due to the particular shaft.

Time averaging appears to be, and is, a very powerful and useful tool for rotating machinery, and for gear drives in particular, but as with all techniques there are liable to be problems or difficulties.

The major problem is associated with "jitter" or "smearing" and is due to variation in speed of rotation. The start of a revolution is given by an accurate 1/rev pulse and with typically 500 data samples per rev the starting position is located consistently within 0.2% of a rev (0.72°). If we have 50 teeth on the gear and are primarily interested in 1/tooth (and harmonics) noise then a 1% variation in speed between one revolution and another would mean that by the end of the revolution two 50/rev waves recorded at the same sampling rate (in time) would have moved 180° in phase relative to one another. If, at the start of the rev they were adding, they would be cancelling each other by the end of the rev. A 1% speed change is unlikely to occur within a revolution or on successive revolutions but might occur over 50 revs which is the sort of order of number of revs over which we might average signals.

Fig. 9.14 indicates the "smearing" effect and shows how the observed averaged amplitude reduces as the signals move out of phase with the speed variation, which in the case shown has given cancellation with 180° phase shift half way round the revolution. The jitter effect causes trouble because we sample at a constant rate in time whereas we wish to sample at constant angular positions round the revolution. On a test rig this could be achieved by fitting an additional rotary encoder (with 512 lines) and sampling the vibration signal when demanded by the next encoder line. This technique is rather too cumbersome for general use.

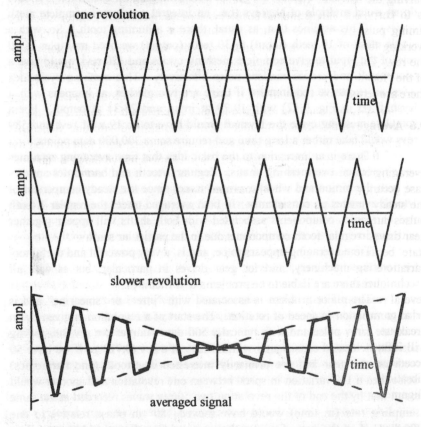

Fig 9.14 Effect of speed variation on time averaged signal.

An alternative to physically fitting an encoder is to work back from when the 1/rev pulses occurred (in time) to exactly when the samples should have been taken (assuming constant speed during the revolution), and then use relatively complex interpolation routines to estimate what the vibration reading would have been if the sample had occurred exactly at the "correct" time. Again this is excessively cumbersome for normal use. It is usually less effort (in total) to restrict data logging to times when the speed remains reasonably steady over a few seconds. An interval counter reading rev times from the 1/rev sensor and set to 10^{-5} seconds resolution is usually useful. An alternative technique is to have two separate sensors, half a revolution apart and average on each separately, using only the first half of each rev and halving the time available to get out of phase. Yet another, better, approach is to carry out the time averaging analysis working backwards from the timing pulses to reduce smearing in the latter half of the rev, as well as working forwards from the pulses to reduce smearing during the first half of the rev. This is relatively easy to do in the analysis routines and comparison of the "forward" and "backward" averages gives a clear indication of whether there are serious jitter problems.

9.6 Average or difference

It is easy to get carried away by the power and usefulness of time averaging but occasionally it is not the average that is important. A classic case occurs with internal combustion engines where we are less interested in the steady firing pulses from the (four) cylinders than in the variation of the pulses due to irregularities in carburation or turbulence. Correspondingly, in gear drives we may be interested in variations of noise pattern from the steady state because human hearing is very sensitive to small modulations or variations from regularity.

Irregular variations in gear drive noise or vibration can occur for several reasons. Intermittent interruptions in oil supply can have some effect, or alignment variations, due to the cage of a rolling bearing beginning to break up, can modulate the signal. External variations due to variable load will influence noise, especially if teeth are allowed to come out of contact, or occasionally dirt or debris passing through the mesh will give transient vibration. Hull twisting on a ship may distort the gear casing and alter alignments of the meshes.

For any problems of this type an effective method is to compute the time averages for both input and output shafts and subtract the averages from the original signal so that what is left is the variation from average or "normal" signal. Care is needed not to subtract the tooth frequency components twice as they appear in both averages. The variation from average can then give an indication of the problem cause, especially if there is an external load or speed variation.

9.7 Band and line filtering and resynthesis

In many vibration signals there are present vibration components, often quite large, which are irrelevant to the investigation. It is of little help to be told that there is a large component at 21 times per revolution if we already know that there are 21 teeth on the gear and that the problem is not at tooth frequency. Similarly a component at mains frequency (or harmonics) is likely to be electrical noise or drive torque fluctuations.

It may be much easier to analyse or assess the time signal if these expected components (which are legitimately present) are removed from the signal. Originally the analog methods employed for this involved either (Fig. 9.15)

(a) notch filters which were often for mains interference, or

(b) "band stop" filters to cut out a range of frequencies, typically an octave.

These helped but were of limited performance and could not deal with any subtleties in the signals. Digital methods are much more powerful and flexible and now dominate the field. Digital filters can give extremely high performance high pass, low pass, band pass or band stop performance to "clean up" a signal by removing known, irrelevant components.

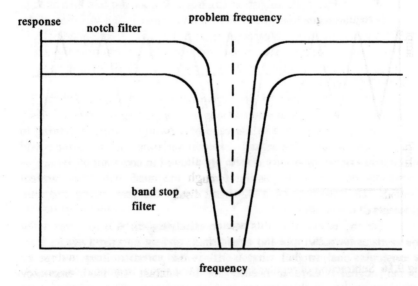

Fig 9.15 Response of analog notch and band stop filters.

They work by convolution, multiplying the raw signal by the filter impulse response, and involve a large number of computations, so it is difficult to achieve high speeds at low cost.

The alternative, line elimination and resynthesis, is based on the standard FFT routines which are fast and efficient and can be easily programmed in Matlab [4] or a similar language. The vibration trace, whether raw signal or rev-averaged signal, is analysed using an FFT routine (a one-line instruction). Then either

(a) known lines which should be there are removed automatically or

(b) the frequency analysis is displayed and the 'legitimate' lines to be eliminated are chosen by the operator or

(c) all lines above a certain (absolute) amplitude are removed (this is the easiest option to program but involves an arbitrary choice of the critical level).

For each frequency in a real signal there are two lines in the FFT because frequencies always appear as conjugate complex pairs in the analysis.

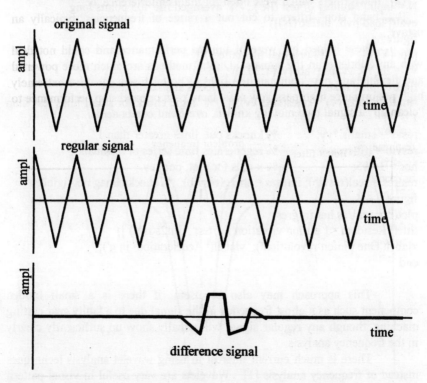

Fig 9.16 Subtraction of regular signal from test signal to emphasise changes.

The selected lines are removed by putting their amplitude to zero. The resulting remaining frequency components are subjected to the inverse Fourier routine (ifft) which resynthesises the original time sequence signal with all the "normal" vibration removed. The residual signal will show up minor faults much more effectively than the original signal.

Fig. 9.16 shows an example of a simple, apparently regular, time signal which has had the regular signal of 1/tooth (and harmonics) subtracted. The difference signal shows very clearly that there was a phase delay (or pitch error) on one tooth in the original signal. The method is especially useful when there are irregularities in small harmonics which cannot be seen due to large components at 1/tooth and similar frequencies.

A typical Matlab program to eliminate the large lines for a once per revolution averaged file obtained in a test is as follows:

```
% loads pinion vibration averaged file pvbN for viewing and line elimination
clear
N = input('Number of test file') ;  %  averaged file 405 points long
eval(['load pvb' int2str(N)]) ;
figure;
plot (Y) ;            % original file called Y
vv = fft(Y); vvabs = abs(vv(1:202)) ;
figure; plot(vvabs);                    % looks at sizes of lines
smalls = (abs(vv) < ones(size(vv)));   %  logic check for small lines less than 1
resvv = smalls.*vv;        % knocks out  lines greater than 1
resvib = ifft(resvv);       % regenerates time series of residuals
hor = 1:405;               % x axis for plot, one rev.
realres = real(resvib); imgres = imag(resvib);  % checks imag negligible
figure
plot(hor,realres,hor,imgres)
title(['Residual <1 pinion vibration for test  ' int2str(N) ])
xlabel('One pinion revolution'); ylabel(' Acceleration  in g');
end
```

This approach may also be useful if there is a small hidden component such as a ghost frequency in the signal due to a faulty gear cutting machine, though any regular signal will usually show up sufficiently clearly in the frequency analysis.

There is much current interest in using wavelet analysis techniques instead of frequency analysis [1]. Wavelets are very useful in visual pattern recognition for detecting sudden steps or transitions such as edges of objects but are less selective when there is steady background vibration. Because gear errors tend to have regular components and faults show up as variations from a regular pattern, the line elimination approach tends to perform better. The advantage of wavelets is their variable time scale but the same effect can be obtained with frequency analysis if corresponding short windows are employed at the higher frequencies. Some of the more sophisticated wavelet shapes look extremely similar to short window Fourier transforms and so give the same results.

9.8 Modulation

A vibration signal may have amplitude or frequency modulation, usually at once per revolution, and this tends to worry operators. The most likely reasons for modulation are:

(a) Variable load torques, especially if the teeth come out of contact for part of the revolution. Alternatively, shaft deflection may vary with load with an overhung gear and modulate the signal as the helix alignment varies. There may also be a small effect due to tooth elastic deflections altering the T.E.

(b) Eccentricities. These may act, usually at 1/rev to vary the torque, and modulate the vibration as in (a).

(c) Movement of the source. This occurs in an epicyclic gear where the planets travel past a sensing accelerometer mounted on the (fixed) annulus. The effect of the different vibration phase on each planet mesh is to produce an apparent higher or lower frequency than the actual tooth meshing frequency. This frequency looks like a sideband of tooth frequency and the tooth frequency itself is often not present [5].

(d) A gear mounted with swash may give a signal modulated at 1/rev or at 2/rev as the alignment of the helices varies.

The modulation is usually amplitude modulation which is easily seen on the original time trace as sketched in Fig. 9.17, but appears as sidebands in the frequency analysis in Fig. 9.18. Not only the basic once-per-tooth frequency but all the harmonics are modulated. In extreme cases the 1/tooth frequency can disappear completely leaving only the two sidebands or occasionally just the single sideband as with an epicyclic drive.

Frequency modulation involves variation of the periodic time of the waveform and cannot be easily seen in the raw signal as the amplitude remains constant (as in Fig. 9.16), but it is easily detected by line elimination. However, the frequency analysis looks almost the same as the result for amplitude modulation (shown in Fig. 9.18).

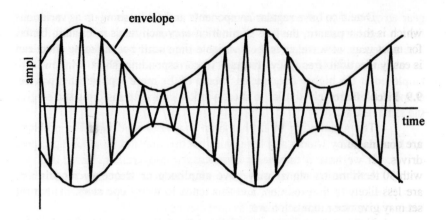

Fig 9.17 Time signal with amplitude modulation.

If it is at low frequency, the modulation may be audible and irritate the customer. Prevention of the torque variation is sometimes not possible, but if the amplitude of the "carrier" (i.e., the 1/tooth) is reduced, the fact that there is modulation will matter less. Eventually if the "carrier" i.e. the tooth frequency component is reduced to zero then there is no sound to irritate the customer.

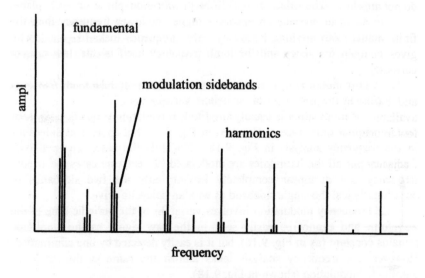

Fig 9.18 Frequency analysis of modulated signal.

Detection of modulation can be assisted by using the "cepstrum" which is the frequency analysis of the frequency analysis, see Randall [2], but for most gear work the effect is clearly visible and the modulating frequency is easily identifiable as a 1/rev frequency.

9.9 Pitch effects

The assumption so far has been that noise and vibration problems are dominated by 1/tooth and harmonics but this may not be so for high speed drives. If we have a turbine or compressor pinion running at 12,000 rpm with 30 teeth the 1/tooth frequency is 6 kHz. In general frequencies this high are less likely to find responsive resonances and give noise problems but the set may give noise at much lower frequencies below 2 kHz.

Noise in this frequency range is at say five times per pinion rev or twenty times per wheel rev and so is rather puzzling. It can be due to phantom or ghost tones from the gear manufacturing machine but such tones are easily identified as they correspond to the number of teeth on the table wormwheel. If not the trouble may be due to random pitch errors on the pinion or wheel.

Adjacent pitch errors are typically of small amplitude and should be rarely larger than 4 µm and as they are random we would expect negligible excitation at any single frequency. The test results may be as in Fig. 9.19 and do not appear to be capable of giving significant trouble.

Although the pitch errors are random in distribution there are only a finite number of teeth round any gear and the sequence then repeats. This gives components of excitation at all possible multiples of 1/rev except curiously at 1/tooth and harmonics of 1/tooth (see Welbourn [6]).

This means that at any multiple of 1/rev (excluding tooth frequency and harmonics) there may be a significant component of that harmonic available to excite structural resonances which are likely to exist at relatively low frequencies.

adjacent pitch error

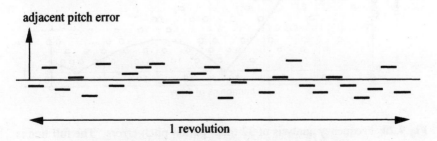

1 revolution

Fig 9.19 Typical adjacent pitch errors around a gear.

The theory gives the result that if very large numbers of gears are tested the average measured amplitude of any given harmonic of order z will be proportional to

$$\sigma\sqrt{\left(\frac{2}{z}\right)}\left[\frac{\sin(m\pi/z)}{m\pi/z}\right]$$

where σ is the rms value of the adjacent pitch errors.

The theory thus predicts that the distribution of harmonics will be as shown in Fig. 9.20 but also predicts that the variations of amplitude in the frequency analysis will be as large as the amplitudes expected on average (the full line). The circles indicate typical measured results which have a large scatter. The harmonic amplitudes expected are surprisingly large.

Taking the original adjacent pitch error as 2 µm rms the expected value of a low harmonic will be as high as $2\sqrt{(2/32)}$ which is 0.5 µm rms or 1.4 µm p-p.

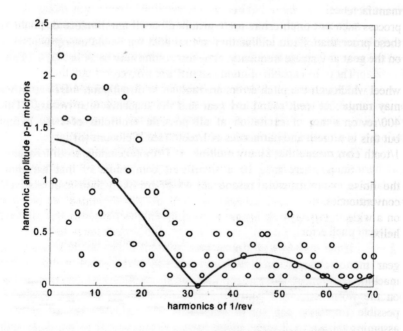

Fig 9.20 Frequency analysis of 32 tooth pinion pitch errors. The full line is the theoretical prediction and the circles are typical experimental values.

On a 5th harmonic this would have dropped to 1.35 mm p-p but any particular gear could easily have over double this value and 3 μm p-p would be likely to give audible trouble.

The other effect that pitch error harmonics can have is to give the illusion of a false phantom note at about 1.5 times tooth frequency. Looking at harmonic 45 gives a predicted amplitude of 0.21 of 0.5 μm rms and so about 0.3 μm p-p with the possibility of double this value, comparable with a phantom on a well made large gear.

9.10 Phantoms

The existence of phantoms was mentioned in section 9.9. They appear in a frequency analysis of noise or T.E. as a "wrong" frequency. It is rather a temptation to ignore them because it seems that if there are 106 teeth on a gear there should not be a vibration at 145 times per rev. Their existence is liable to be blamed on some unknown electrical interference or sampling frequency fault. They may however be genuine.

They are normally caused by the machine on which the gear was manufactured, whether a hobber or grinding machine. Even though a final process such as honing, shaving or grinding may not in itself cause phantoms these processes tend to follow the previous pitching so that any problems left on the gear at the roughing stage may not be eliminated in finishing.

They are usually caused by the 1/tooth error from the worm and wheel which is the final drive to the table carrying the gear and the frequency may range from 90/rev typically on a small machine to between 300 and 400/rev on a large machine. Amplitudes are small, of the order of 1 to 2 μm but this is more than sufficient to be audible and is sometimes larger than the 1/tooth component.

Such phantoms or ghost tones in a gear are clear and consistent in the noise, vibration and in the T.E. They are not easily detected by conventional profile or pitch checking but it is sometimes possible to see them on a wide facewidth gear in the helix check as they appear as a wave on the helix.

If the existence of a phantom throws suspicion on the accuracy of a gear manufacturing machine it is relatively straightforward to test the machine table accuracy directly. One encoder mounted on the table and one on the worm drive shaft give the T.E. directly and it is then sometimes possible to adjust the worm alignment to minimise the 1/tooth error, assuming the worm has been mounted in double eccentric adjustable bearings to allow adjustment of clearance and alignment.

Another hazard that can be encountered is a torsional vibration linked to the revolution of a pinion appearing to be 1/tooth or a modulated

1/tooth but caused by a driving stepper motor. Stepper motors are popular drives for positioning due to the simplification of the control aspects but have the disadvantage that they cannot accelerate high inertias. The designs must ensure that the moment of inertia seen by the motor is small and there is then a possibility that the steps of the motor will insert torsional vibration which, in extreme cases, can reverse motor direction each step allowing gears to come out of contact.

References

1.   Newland, D.E.N.,   'Random vibrations, spectral and wavelet analysis.' Longman, Harlow, UK and Wiley, New York, 1993.
2.   Randall, R.B., 'Frequency analysis.' Bruel & Kjaer, Naerum, Denmark, 1987.
3.   Schuchman, L.,   'Dither signals and their effect on quantization noise'. IEEE Transactions on Communications, Vol. COM-12, Dec.1964, pp 162-165.
4.   The Math Works Inc., Matlab, Cambridge Control, Jeffrys Building, Cowley Road, Cambridge CB4 4WS or 24 Prime Park Way, Natick, Massachusetts 01760.
5.   McFadden, P.D. and Smith, J.D., 'An Explanation for the Asymmetry of the Modulation Sidebands about Tooth Meshing Frequency in Epicyclic Gear Vibration.' Proc. Inst. Mech. Eng., 1985, Vol. 199, No. C1, pp 65-70.
6   Welbourn, D.B., 'Forcing Frequencies due to Gears.' Conf. on Vibration in Rotating Systems, I. Mech. E., Feb. 1972, p 25.

10

Improvements

10.1 Economics

Returning to the basic ideas of noise generation we have:

Gear Errors, Deflections, Distortions, etc.

giving

Transmission Error

which acts on internal dynamics

giving

Gear Body Vibration

and hence

Bearing Housing Forces

which excite the gearcase or transmit through feet

giving

Panel Vibrations

and hence

Noise.

We can (in theory at least) improve any part of this chain and the end result, in a linear system, will be less noise. Hence, we have the choice of tackling (and improving) the transmission error, the internal dynamic response, the external structure dynamic response, or the sound after it is out of the metal.

Once the initial investigations have been carried out the choice must be made as to where improvements should be tried. In general, the choice must (or should) be dictated by economics, economics or economics.

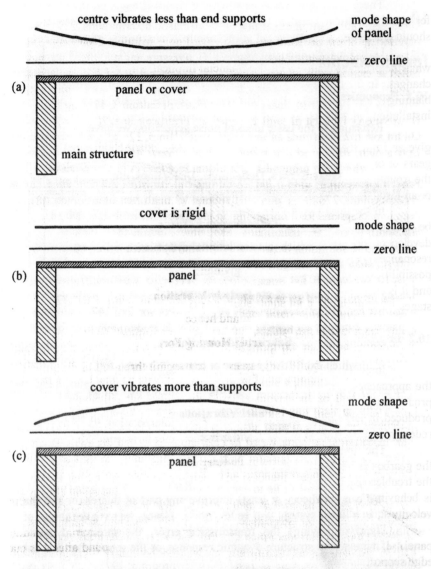

Fig 10.1 Vibrating shapes of panels.

This usually rules out tackling the sound after it has left metal. Absorbing sound without an airtight enclosure is difficult and preventing air circulation does not help cooling.

There are a few occasions when the choice is made on time scale or for purely political reasons but for the majority of problems, economics should dominate.

Unfortunately this means having a rather good understanding of what the problem is and what the financial implications are of a given set of changes. In the middle of a high adrenaline situation with installation design blaming "lousy gears" and the gear production blaming a "hopeless installation," this is not always easy and sometimes impossible.

The dominating requirement is to determine the T.E. since this will give an immediate clue as to whether the problem can be attributed to poor gears or an over-sensitive installation. Without knowledge of the source of the trouble much money can be wasted on attempting to improve a gear pair or an installation that is already extremely good.

In the limit the problem may be so intractable that every aspect must be improved. Fortunately this is rare and only occurs when several developers have already had a go at improving the installation stiffnesses, resonances, and gear design details and have eliminated all the easy possibilities. As often in engineering there is a law of diminishing returns and it is only possible to get dramatic 10 dB or 15 dB reductions in the initial stages.

10.2 Improving the structure

Improving the structure is usually the simplest and most obvious of the approaches. It is generally not the most economic approach for a 1-off production problem but is by far the most economic for anything that is being produced in large quantities. Any improvement is gained with some initial redesign cost but little subsequent cost per item.

The first move is to run round the gearcase (or machinery in which the gearbox is installed) with an accelerometer feeding into an analyser set to the troublesome frequency. The hope is to find some large, flat panel which is behaving as a very good loudspeaker. The relevant criterion is roughly velocity squared times area of panel for sound emission [1].

Fig. 10.1 shows sketches of possible mode shapes for a cover or panel. If vibration amplitudes measured in the centre are greater than the edge support amplitudes [10.1(c)] the panel is acting as a loudspeaker (at the relevant frequency). If panel centre vibration amplitudes are less than edge support amplitudes [10.1(a)] the cover is giving less sound than would a perfectly rigid cover [10.1(b)] so it should be left strictly alone. It is sometimes possible to isolate a panel completely from its support but this is not common.

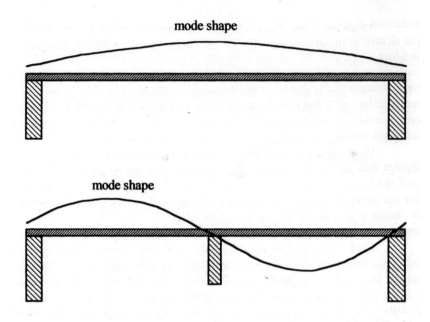

Fig 10.2 Effect of centre rib on mode shape for a vibrating panel.

Individual "amplifying" covers or panels can have their sound transmission greatly reduced either by thickening the panel or by adding a stiffening rib in the centre. Fig. 10.2 illustrates the difference in mode shape between a panel with an effective centre rib and one without.

Technically, the centre rib restricts movement so that the 2 half panels can only vibrate in anti phase (as a dipole) and their emitted sound waves (180 degrees out of phase) tend to cancel, once they are well away from the panel. The rib has to be quite deep to be effective on a flat cover and, within a casting or weldment, it helps if an internal rib is also taken across the corner onto a neighbouring panel. The resonant frequencies of the panel are greatly increased.

Gearcases which are cast tend to be much quieter than the corresponding weldments. This is not, as customarily assumed, because cast iron has greater damping than steel because both have very small damping in absolute terms. The main reasons for the difference are that curved cast surfaces are much more rigid than flat surfaces and, because iron casters are paid by weight, castings are usually much thicker than the corresponding weldments. As plate bending stiffness is proportional to thickness cubed, this provides a major increase in rigidity despite the lower modulus of elasticity. There is also likely to be an increase in corner stiffnesses and an effective reduction in span due to the radii associated with casting. It is of interest that the structural rigidity of a weldment in torsion is little affected by the depth of welding at the corners. In a normal gearcase, stresses are negligible because high stresses would give ridiculous movements so it is not necessary to have high strength at the welds. This means that within a given cost, it is often much better, from the structural and noise aspects, to have thick panels with only (unchamfered) fillet welds rather than thinner panels with (expensive) full depth welds.

If all the individual panels have already been stiffened and split into dipoles then little can be done without a major increase in weight. Increasing wall thicknesses gives major stiffness increases (but with weight penalties) but use of aluminium or magnesium alloy panels allows large increases in thickness and hence plate bending stiffness without weight penalties (but at a cost).

Cars and office machinery have a problem because there are large thin flat panels. On a car it is not possible to increase panel thickness due to weight penalties and although improvements can be made by adding highly viscous bitumen-based damping pads on the panels there is, again, a weight penalty. Modern body designs tend to have more curved panels, not because of styling considerations but as an aid to increased stiffness. The ideal structural shape is a sphere. Office machinery traditionally has flat panels so great care has to go into isolating the drives from the panels. Plastic may be used to increase wall thicknesses and, hence, rigidity and damping, despite the low modulus of plastics.

At the design stage there will not be a structure available to test but occasionally there is a smaller but similar gearbox available. Once the smaller gearbox has been tested the natural frequencies of the larger design can be estimated. The relevant non-dimensional parameter for natural frequency is $\omega^2 L^2 \rho / E$ so since the material is the same, the product of natural frequency and size should remain constant. Typically a 25% increase in all dimensions should give a 20% reduction in natural frequencies provided geometric similarity is maintained. The existing gearbox can then be tested at 125% speed to give an idea of the vibration responses to be expected.

10.3 Improving the isolation

Most machinery has the gearbox isolated from the main structure by rubber mounts. If not, the design is asking for noise troubles. Unfortunately, the isolation mounts have very rarely been designed with the specific intention of isolating the 1/tooth frequency which is usually the main excitation. Sometimes, as in an elevator drive, it is difficult to isolate the drive from the customer (in the lift cage).

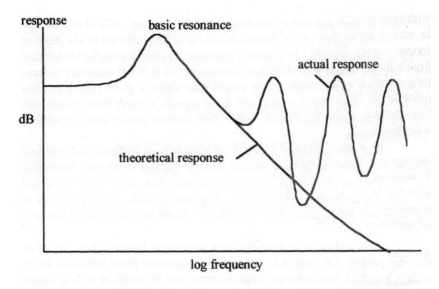

Fig 10.3 Typical response of vibration isolator.

Many installations have isolators which were designed to isolate 1/rev (often 1450 rpm, 24.5 Hz) and simple theory says that the isolation should then be very good at 24/rev (i.e., tooth frequency of 600 Hz). Fig. 10.3 shows the theoretical single degree of freedom response and what may realistically happen as the internal resonances of the spring give "spring surge," the bane of racing engine valve springs.

Satisfactory isolation of tooth frequency needs a design tailored to tooth frequency, so either the isolator should be redesigned for the higher frequency, or two stage isolation is needed when both 1/rev and tooth frequency are involved. The 1/rev will not come through as noise because frequencies are too low but will be felt as vibration whereas 1/tooth noise frequencies cannot usually be felt as vibrations. As with all 3-dimensional isolation it is important that lateral or vertical vibration and torsional vibration modes are decoupled to prevent interactions. This is most important in a car where there are large torsional vibrations of the engine, especially at idling. If these were allowed to interact to give vertical body movement, there would be severe passenger irritation.

Another problem comes from large "static" loads. We need relatively soft support springs to give good vibration isolation but if high average loads are imposed, the springs must be stiff to prevent excessive geardrive movement. This problem occurs in cars because with a transverse mounted engine, gearbox, and differential assembly, the system must withstand reaction torques of the order of 2000 Nm (1500 lb ft) at full throttle in bottom gear but it must be quiet when cruising on a motorway when the torque is only 100 Nm (75 lb ft). The most satisfactory solution is to have a highly non-linear support which is soft at low torques and locks up when the torque rises (see section 6.5). Fortunately, a driver is not worried about high noise levels for a couple of seconds at full throttle in lower gears when the high torque involved "bottoms" the support and there is high vibration transmission.

In a very sophisticated installation the "ultimate" isolation is to indulge in vibration cancellation techniques at the (four) gearbox support feet in addition to using soft mounts. This is technically easier than cancelling airborne sound after it has escaped from the metal. It is, however, a very expensive, delicate and temperamental method which should be avoided for all normal engineering.

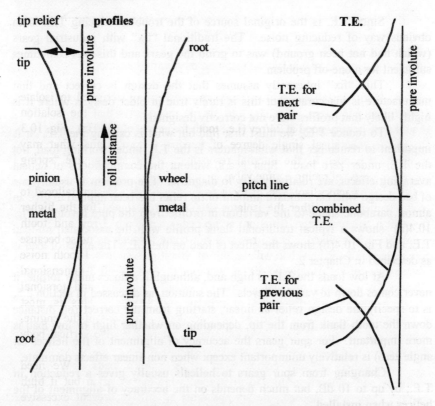

Fig 10.4 (a) Flank profile shapes combining to give T.E.

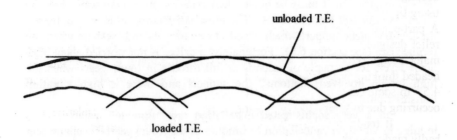

Fig 10.4 (b) Effect of load on T.E for a spur pair.

10.4 Reducing the T.E.

Since T.E. is the original source of the trouble, reducing T.E. is an obvious way of reducing noise. The traditional "fix" with industrial gears (which had not been ground) was to grind the gears and this was sometimes sufficient for a one-off problem.

This "fix" inherently assumes that the design is correct and that manufacture is inaccurate, but this is rarely true in older designs where it is highly likely that profiles were not correctly designed.

To reduce T.E., we must first find out what is causing the T.E. It is important to remember that what matters is the T.E. under working load, not the T.E. under zero load. Spur gears, without the complications of helical averaging effects, are relatively easy to diagnose. The problem is usually one of bad design where a standard amount of tip relief has been applied to give an almost parabolic shape to the variation in profile from the pure involute. Fig. 10.4(a) shows a typical traditional flank profile with the associated no-load T.E. and Fig. 10.4(b) shows the effect of load on the T.E. The effect of load is as described in Chapter 2.

At low loads the T.E. is high and, although it reduces under torque, it never comes down to very low levels. The solution, as discussed in section 2.5, is to specify the design relief as linear, starting from the correct roll distance down the tooth flank from the tip, depending on whether high or low load is more important. For spur gears the accuracy of alignment of the helices (at angle zero) is relatively unimportant except when non-linear effects dominate.

Changing from spur gears to helicals usually gives a reduction in T.E., by up to 10 dB, but much depends on the accuracy of alignment of the helices when installed.

Reducing the T.E. on helical gears is a much more difficult process due to the complex interaction between helix and profile effects. Much depends on whether the original design attempted to achieve a smooth entry by using tip relief and negligible end relief, or end relief with negligible tip relief. A particular case of the latter occurs with heavily crowned gears with no tip relief designed for light loads. Improving gears where there are no obvious major design errors will usually involve either an amount of extremely clear-headed thinking or the use of at least a thin-slice model as described in section 4.5. In some cases the dominant effect can be the variation in helix matching occurring due to shaft deflections under load.

It tends to be assumed that gears are noisy because they have been badly made and there is the inherent assumption that the gears will have been well designed, usually the exact opposite of reality. Gears are often manufactured to within 3 μm of the design profile specification which itself may be 15 μm in error. For any old design it is well worth checking the levels of T.E. that would be predicted from the specified tooth shapes. In any prediction it is important to feed in some helix errors since a perfect helix match will often give low T.E. regardless of profile shape, but perfect helix matching is unrealistic. Even in a modern design it is worth checking that long relief has not been used instead of short relief or vice versa. Although much can be deduced from design drawings, there is no substitute for experimental measurement of the T.E.

10.5 Permissible T.E. levels

Inevitably, in a development or problem investigation the question will arise "what is the permissible/correct/reasonable level of T.E.?" Specifications (DIN and ISO) for once, are of no use whatsoever, partly because even when they reluctantly mention T.E. they do not correctly specify the parameters that are relevant for noise purposes with sufficient care (Fig. 10.5).

F_1' and f_1' [2,3] are in themselves no help since, for noise purposes, the eccentricity effects which dominate F_1' are almost completely irrelevant and we are interested in the semi-steady 1/tooth component and harmonics, not an odd peak f_1' value read off a curve which has been distorted by eccentricity. In addition, there is no general pool of knowledge in industry as to what level may or may not be suitable. To get a sensible value for the 1/tooth error it is necessary either to carry out a frequency analysis or at least to filter out 1/rev effects.

The ultimate control on T.E. is what the customer will tolerate in that particular installation. There have been many instances where a gearbox was perfectly satisfactory in one car but sounded terrible in a different model. In any industry it is almost inevitable that a manufacturer will have to cross-check T.E. against final installed noise.

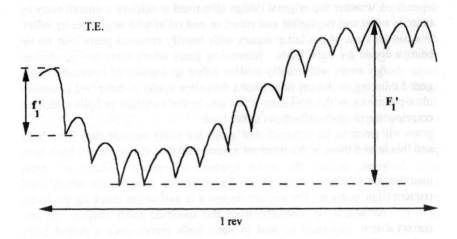

1 rev

Fig 10.5 Typical T.E. showing how eccentricity gives false 1/tooth error.

This cross-check is partly to convince everyone that the two are connected but mainly to set permissible levels on T.E. This may result in some major variations in that a car gearbox may require loaded T.E. to be less than 3 μm in 5th, 5 μm in 4th, 2 μm in 3rd (because of a particular difficult resonance), 7 μm in 2nd and 12 μm in 1st gear. It is worth noting that when permissible T.E. is quoted, it is necessary to be extremely legalistic and to specify whether it is peak-to-peak of total 1/tooth and higher harmonics (cutting out eccentricities only) or peak-to-peak of 1/tooth (filtered) or peak of 1/tooth or rms of 1/tooth.

When the signal is modulated there are even more possibilities according to whether maximum or average values are taken during a revolution.

An industrial general purpose gearbox will be used in many installations, some good and some bad so we need a "reasonable" T.E. level setting which is independent of installation. As mentioned in section 10.1 we can then target either gearbox or installation according to whether the measured T.E. is above or below the "reasonable" level.

A "reasonable" level of T.E. depends on price and it is unrealistic to expect an "industrial" cheap gearbox to attain the same T.E. figures as one costing three times as much, although cost and quietness are not always linked. Curiously the levels of T.E. (in μm) are roughly independent of gear size so diameter is not a major variable. It is difficult to convince gear users that a well made 4 mm diameter gear is liable to have the same absolute size errors as a well made 4 m diameter gear but this, surprisingly, is reality.

The starting point is rather arbitrary, but fortunately in the S.I. system there is a convenient "round amount" at about the right point for demarcation. We can take a figure of 10 μm (0.4 mil) peak-to-peak at 1/tooth frequency as being a dividing line between rough and very poor gears.

A T.E. of 20 μm p-p would only be permissible on a large slow-speed gear for the sort of machinery where gear noise is not really a problem. At the ultra-precision end, a T.E. of 1 μm p-p is extremely good and is correspondingly very rarely achieved. Medium and small sized industrial gears will generally be very satisfactory with less than 3 to 4 μm at 1/tooth p-p and this level should be achieved with quality gears.

It should be noted that these are "loaded" values and values on a no-load test for spur gears will generally be higher so that, under load, the T.E. reduces (if properly designed).

Another factor which should be checked is whether the T.E. is the correct shape. In Fig. 10.6, curve A is what we would expect from a spur gear and curve B is typical for a helical gear.

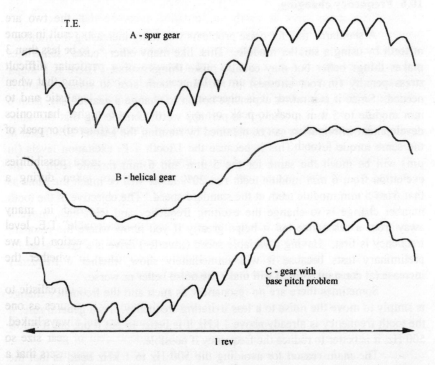

Fig 10.6 Different once-per-tooth T.E. shapes with spur, helical and faulty gear.

Curve C suggests that something has gone badly wrong with the geometry since the sudden drops at the ends of each tooth pair suggests that the base pitches on the two gears are not the same. This may be due to an incorrect design lean or correction on an involute profile or just due to bad manufacture.

T.E. figures of less than 5 μm p-p with desired levels of perhaps 2 μm appear to be extremely accurate by normal metrology standards especially as we normally need to measure a factor of 10 more accurately to meet specifications reliably. There is, however, no problem in achieving this accuracy of measurement reliably and consistently with single flank checkers. The accuracy figure is relevant to information which has been frequency analysed, giving large improvements in accuracy because the accuracy at tooth frequency is typically a factor cf 30 better than the quoted encoder accuracy [4]. Measuring accuracies of 0.1 μm are easily achieved provided there is no dirt or airborne dust on the tooth flanks but the average metrology shop is not a clean room.

10.6 Frequency changing

A standard "fix" for noise problems was to try increasing the numbers of teeth by using a smaller module. This, like many other "cures" sometimes makes things better but may equally make things worse. There is a major stress penalty (in root stresses) in reducing tooth size so some caution is needed. Since it is a rather expensive option to change a gearset, say, from 6 mm module to 5 mm module, it is perhaps worth considering that the same development information can be obtained by running the gearset 20% faster at the same torque level. This is because the 1/tooth T.E. excitation levels (in μm) will be much the same for the 5 mm and 6 mm module teeth so the excitation from 6 mm module teeth run 20% faster will be much the same as that from 5 mm module teeth at the standard speed. The objective of the tooth number change is to change the exciting frequency and with luck, move it away from a resonance but it helps greatly if you know where the resonant frequency is first. Having a variable speed (inverter) drive is a great asset for preliminary tests because it will immediately show whether a frequency increase (at constant torque) will make the noise better or worse.

Sometimes there are no resonances as such and the frequency change is simply to move the noise to a less irritating frequency. As a general rule, if the tooth frequency is already above 1 kHz it is better to put it up, but if below 500 Hz, it is better to reduce the frequency if possible.

The main reason for avoiding the 500 Hz to 1 kHz band is that the human (A-weighted) ear is most sensitive in this range and also because many structures are at their noisiest in this range. At high frequencies the wavelengths are smaller and panel vibrations have a greater tendency to be in anti-phase and cancel. At low frequencies, velocities and, hence, noise pressure levels drop and also hearing sensitivity drops.

10.7 Damping

It is tempting to think that it should be possible to introduce damping to reduce noise levels, either inside the gearbox or in the structure of the installation.

Damping of very thin panels such as car body panels is successful in reducing vibration and noise levels, but attempts to increase damping in a gearcase are not usually very successful. Adding a pad of viscoelastic material to a car panel 0.75 mm (30 mil) thick can absorb a high proportion of the bending wave energy passing through the panel and natural frequencies are reduced due to the extra mass but if the "panel" is 1" (25 mm) thick steel there are no suitable materials to extract much energy. Machine tool designers have attempted to insert damping layers at interfaces between castings, but this approach has not been successful and the use of materials such as synthetic granite, though having nominally higher damping than steel, sacrifices stiffness.

An approach which has been successful in unstressed components such as internal combustion engine rocker box covers has been to sandwich a damping layer between two aluminium alloy sheets.

Scaling this up to industrial gearbox thicknesses does not appear to work although some large gearboxes have used a layer of sand between two steel skins. Whether the principal effect of the sand comes from its mass, from its damping, or from its action in spacing the steel panels apart, has not been stated.

As previously mentioned, although cast iron has higher damping than steel, the effect of the material damping is negligible compared with the damping from bolted joints, shrink fits, loose members rattling about and energy being dissipated into foundations. When plastic casings were used for domestic kitchen equipment the noise levels have tended to be higher than for the previous cast metal casings despite the higher internal damping of the more flexible plastic.

The one technique that has been used over a wide range of industries, with reasonable success, is the tuned damped absorber. An auxiliary mass is supported on a damped spring which is usually deformable nitrile or butyl rubber and is tuned to just less than the frequency of the troublesome resonance. Nitrile rubber is popular as it has nearly the optimum level of internal damping, even at very low amplitudes of vibration. The theory was worked out by Den Hartog over 60 years ago (Fig. 10.7) [5].

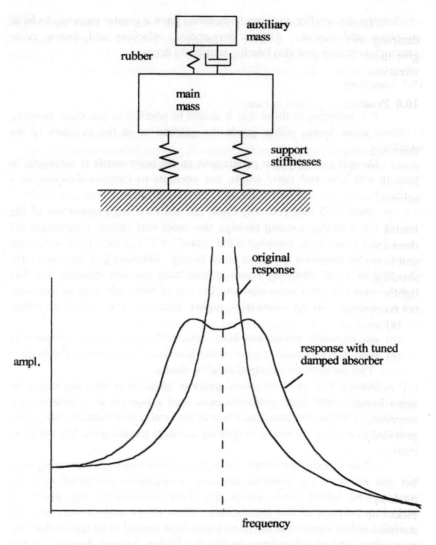

Fig 10.7 Tuned damped vibration absorber response.

    Although it is possible in theory to use steel springs and oil damping, this is rare due to sealing and tuning problems.

    The device needs careful tuning to the correct frequency and is, in general, only worthwhile if the auxiliary mass can be about 10% of the effective mass of the resonance and the original dynamic amplification factor (Q) of the resonance was greater than 8. The absorber can then reduce the Q factor to below 4.

Untuned (Lanchester) dampers which use only mass and viscous damping will work over a range of frequencies but require greater mass and give much less damping so they are little used except for torsional engine vibrations which occur over a wide range of frequencies as speed varies.

10.8 Production control options

When trouble strikes and the customer's installation cannot be altered there is a tendency to panic and to halve all drawing tolerances on principle, to make sure that all the gears are being made "better." This is, of course, no help if it is a faulty gear design (or installation) and is very expensive to achieve.

On the assumption that development has investigated permissible loaded T.E. and found that it must be kept below, say, 4 μm at once-per-tooth, there are several options available. The first is the obvious one to run a model and to see how tolerant the design is to errors of profile, helix and pitch. This should give a good idea of the sensitivity of the design which could decide how tightly manufacturing tolerances should be specified. If these tolerances are not economically sensible then the choices are:

(a) alter the design to make it less sensitive (if possible);
(b) greatly reduce tolerances; or
(c) manufacture scrap.

Option (b), though often used, is usually far too expensive. Option (c), deliberately catering for a percentage of scrap, is guaranteed to produce acute hysteria with production directors and accountants. However, it is surprisingly often the most economic solution and is politically permissible provided that the small percentage of noisy boxes are not allowed to go to the customer. This means 100% T.E. checking on the production line.

This suggestion of 100% T.E. production checking seems expensive but may actually save money because some of the earlier checks on profile and pitch can be reduced or eliminated since detailed faults or changes will be picked up by the T.E. check. There is also a large hidden bonus, due to the statistics of the process, provided that a pair of mating gears are checked as a pair, not separately against "master" gears which these days may well be little more accurate than the gears they are meant to be testing.

If gears are checked individually for a total error band of 4 μm in the mesh then each gear must individually be within +/- 2 μm to ensure that any pair are within 4 μm. This could well generate scrap rates of the order of 10% on wheel and pinion.

Testing together will greatly reduce the scrap rate, as indicated in Fig. 10.8 since, of the "scrap" pinions, most of those "negative" will encounter wheels which are not too large and will mate satisfactorily.

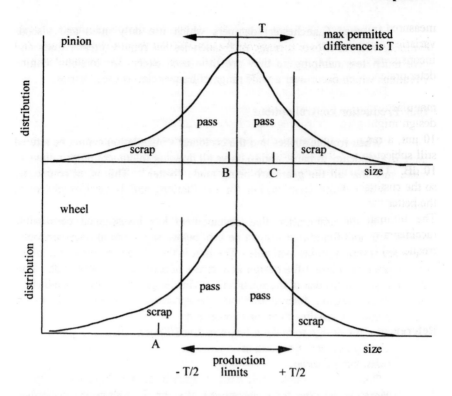

Fig 10.8 Combination of tolerance limits with gear pair testing showing how the number of failed gears is greatly reduced.

A wheel of size A (which must be scrapped if tested separately) will mate perfectly with a pinion of size B, and with any pinion of a size less than C, covering about 75% of the pinions manufactured. This effect can easily reduce scrap rates by a factor of four with corresponding savings.

The cost of T.E. checking is relatively low. The standard commercial checker can cost up to $300,000 (£200,000), much the same as a profile, helix or pitch checker but the testing is very fast (it can easily be < 1 minute) so throughput is high, reducing costs.

Alternatively, a dedicated check rig can be set up for a standard component such as a back axle. The cost of the mechanics, encoders and electronics is then of the order of $30,000 (£20,000) since all the high precision slides and variable settings of the general purpose equipment are not needed.

There is one hazard which sometimes causes puzzlement when gear design is improved and that is the oddity that the statistical scatter on the final noise levels is increased. A poor and rather noisy design might give a measured noise level variation of ± 2 dB. When the design is improved, the variation can easily rise to ± 5 dB so the customer may complain about greater inconsistency in the gear noise and assume that quality control has deteriorated.

The reason for this is that the variations in T.E. are mainly due to manufacturing so they will stay roughly constant at, say, ± 2 μm. A poor design might give a fairly regular "design" T.E. of 8 μm so ±2 μm gives 6 to 10 μm, a range of roughly 4 dB. Improvement of the average T.E. to 4 μm, still subject to ± 2 μm variation gives a range of 2 to 6 μm or a total range of 10 dB. This manufacturing range cannot be reduced by the improved design so the customer has to be educated. It is difficult to convince a customer that the better the basic design, the larger the statistical variation will appear to be. The ultimate case is when the design is good enough to occasionally (accidentally/miraculously) give zero T.E. and the dB range (at a given frequency) is then infinite, regardless of how quiet the average gear pair is.

References

1.   Fahy, F.J. Sound and Structural Vibration. Academic Press, London, 1993.
2.   Maag Gear Handbook, Maag, Zurich, 1990 (in English), section 5.271.
3.   DIN 3963, Tolerances for cylindrical gear teeth. (in English), DIN standards, Beuth Verlag GmbH, Berlin 30.
4.   Smith, J.D., 'Gear Transmission Error Accuracy with Small Rotary Encoders,' Proc. Inst. Mech. Eng., Vol. 201, No. C2, 1987, pp 133-135.
5.   Den Hartog, J.P., 'Mechanical Vibrations.', Dover, New York, 1985, Section 3.3.

11

Lightly Loaded Gears

11.1 Measurement problems

The first hint that a gear drive may be "lightly loaded" usually comes when vibration or noise measurements do not make sense. Amplitudes vary for no apparent reason, frequencies appear which bear no relation to tooth frequency or the "phantom" frequency (from the gear manufacturing machine) and, most characteristic of all, the vibration levels are extremely dependent on load levels.

The standard response of taking a test run and doing an FFT analysis just produces even more confusion as the signal gives roughly equal amplitudes at all frequencies and appears to be trying to approximate to white noise. There may be stronger components near tooth frequency and harmonics but there is a high background continuous spectrum right through the range. Even worse, there may be significant peaks at half tooth frequency and half phantom frequency or at other subharmonics of the obvious frequencies, or at curious ratios such as two-thirds of the tooth meshing frequency.

Since all the rules of linear vibration are being broken, the obvious deduction is that the vibration is non-linear and that application of intelligence rather than mathematics may be required. Since all frequency analysis is based on the assumption of linearity, it is hardly surprising that non-linear systems cause trouble since most vibration engineers have been brainwashed (at university) into carrying out an FFT before they start thinking.

The first question usually asked is "what do you mean by lightly loaded?" This is best answered by saying that when the angular accelerations of the system multiplied by the effective moment of inertia exceed the steady load torque, which is trying to keep the teeth together, then the teeth will start losing contact since the dynamic component is greater than the mean torque level.

This can occur when the angular accelerations (due to T.E. or torsional vibration) are high, the moment of inertia is high or the load torque is low. This is analogous to driving very fast over a bumpy road when (above a critical speed) a lightly loaded trailer will start leaving the ground.

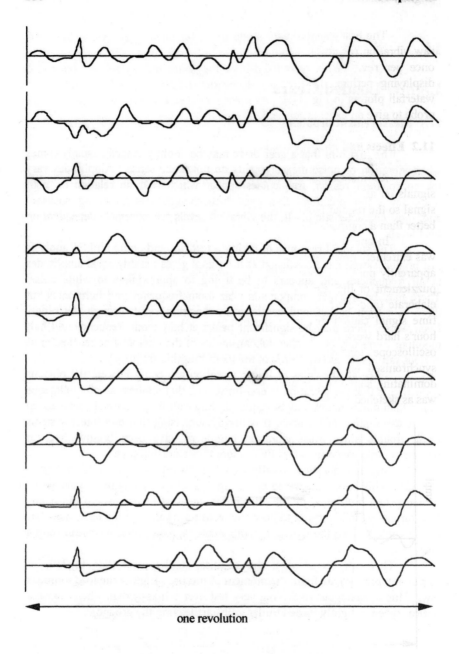

one revolution

Fig 11.1 Vibration on successive revolutions of gear.

The first essential with a non-linear (or linear) system is to look at the raw vibration (or noise) signal on the oscilloscope, preferably synchronised to once per rev. With recorded traces the same effect can be obtained by displaying perhaps 10 revs in succession staggered down the page like a waterfall plot as in Fig. 11.1. As always it is very worthwhile having a 1/rev probe to give an exact synchronising signal.

11.2 Effects and identification

As mentioned previously, humans are good at averaging viewed signals on an oscilloscope or the same effect comes from time averaging the signal so the regular part of the pattern can be seen. In many cases a human is better than a computer for seeing what is happening.

In one engine test in an anechoic chamber, at idling, the timing train was extremely noisy and FFT analysis of the output from a microphone gave apparently pure white noise with no individual frequency peaks, much to the puzzlement of the team of development engineers. The installation was so elaborate (and extremely expensive) that a request for a look at the original time signal caused dismay because it was not available. However, after an hour's hard work the relevant signal was located and brought out to a simple oscilloscope together with a 1/rev pulse. Once the signal had been synchronised on the display, no explanatory words were needed and the dominating sound was of heads being banged against walls. The time signal was as sketched in Fig. 11.2.

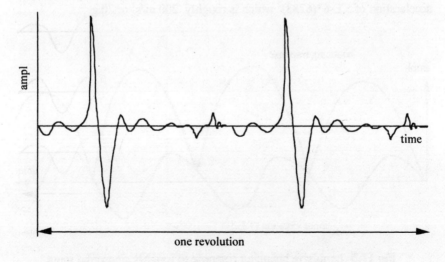

Fig 11.2 Time trace of vibration synchronised to once per rev.

The time signal not only showed clearly what was happening in this case but showed exactly where in the revolution the large engine torsionals were acting to bring the timing gear teeth back into contact impulsively. The fundamental frequency, 2/rev, about 25 Hz, was too low to be picked up powerfully by microphone or accelerometer so it was solely the high harmonics (with much modulation) that dominated the measurements. As far as frequency analysis is concerned there is no difference between amplitude distributions for white noise and for isolated short impulses (see section 9.3).

Both distributions contain equal amplitudes at all frequencies and the only difference is in the phase synchronisation at the pulse.

More commonly, the torsional excitation is due to the T.E. so there is a likelihood of an impulsive vibration at about 1/tooth frequency, varying in amplitude and period. The mechanism (Fig. 11.3) is similar to bouncing a ball on a tennis racket or driving over a very bumpy road at high speed. A short and rather violent impact is followed by a "flight" out of contact until the load torque (or gravity) brings the teeth back into contact after about one cycle of T.E. excitation. It is perfectly possible to bounce powerfully enough to land 2 or 3 cycles later and we then have the "subharmonic" phenomenon of an excitation at 1/tooth giving an irregular vibration at once per 2 teeth or once per 3 teeth. It is difficult for the bounce to maintain consistent time and this gives a very irregular variation in bounce height.

It may seem strange that an excitation as small as T.E. can give trouble, but feeding in a few typical figures shows what is involved. A T.E. of ± 5 μm (0.2 mil) at a 1/tooth frequency of 1000 Hz corresponds to an acceleration of $5 \text{ E-6} * (6283)^2$ which is roughly 200 m/s^2 or 20 g.

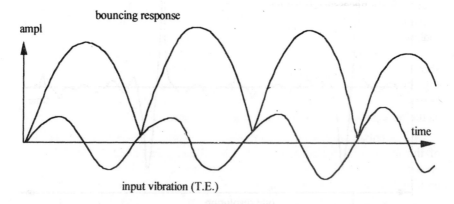

input vibration (T.E.)

Fig 11.3 Impulsive bouncing response to roughly sinusoidal input.

A pinion of mass 20 kg will have an effective linear mass J/r^2 at pitch radius of about 10 kg so to keep the teeth in contact requires a load of about 2000 N (450 lbf) which at 0.1m radius is 200 N m (150 lb ft). This is easily achieved in a normal loaded gearbox but, in a machine such as a printing machine, 20 g acceleration on a printing roll with an effective mass of 500 kg would require 10 tons tooth load, and the load due to printing is at least an order lower than this, so it is difficult to keep teeth in contact.

Testing with portable high speed T.E. equipment on a printing machine will show the manufacturing gear errors repeating consistently at low speeds but as the speed rises the observed T.E. becomes erratic and the drive can be seen bouncing out of contact for long periods.

From an understanding of the basic mechanism it is soon clear that varying the load on the system will have a major effect on the vibration and the quickest and most telling test for non-linearity is to vary the load. This may mean temporarily braking the driven component to increase the torque despite the power waste involved. Major changes in vibration immediately indicate non-linearity whereas minor (<30%) changes suggest a linear system. Curiously, both increasing and decreasing the load may make the system better. If the vibration becomes worse, then usually the alternative will improve it.

11.3 Simple predictions

As with all problems it helps to have a simple model of what is happening to see what the effects of varying the parameters are likely to be. The methods using a full computer time-marching approach as described in chapter 5 are necessary if we wish to detail the effects of misalignment, profile, crowning, etc., in a multi-degree of freedom system. Simple systems can be looked at rather quickly by making some very basic assumptions.

The simplest possible model is the single degree of freedom system shown in Fig. 11.4. The response of this system will have the shape shown in Fig. 11.5. The torsional moment of inertia has been turned into an equivalent "linear" mass. Due to the non-linearity, any original narrow resonance widens as the resonance bends to the left at high amplitude.

Contact will be lost initially when $F = my\omega^2$, where y is the vibration of the mass. The response above this frequency is generally unstable and erratic but we can make some estimates for the condition of maximum amplitude just before the downward jump.

We make the assumption that there are no energy losses during the "flight" so that the initial "upward" velocity is the same as the final "downward" velocity.

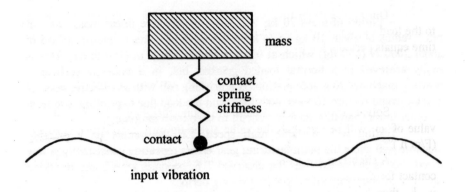

Fig 11.4 Simple model of non-linear system.

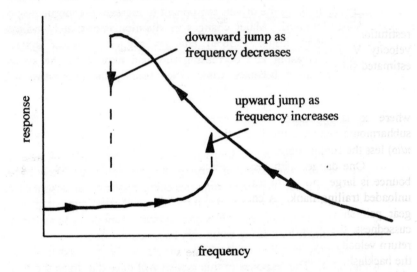

Fig 11.5 Response of "bouncing" system as frequency varies.

Taking the coefficient of restitution at the short impact as e and the "landing" velocity as V then, as the maximum upward velocity of the "base" is hω (where h is the amplitude of vibration of the base), the relative velocity after impact must be e times the relative velocity before impact:

$$(V - h\omega) = e (V + h\omega)$$

During the flight time there will be a constant restoring force F due to the load torque so the acceleration downwards will be F/m and, since flight time equals periodic time

$$2Vm/F = 2\pi/\omega$$

Solving gives ω and V and the bounce height will be $mV^2/2F$. The value of ω will be less than the value at the upward jump which is roughly $(F/m\,h)^{0.5}$.

A slightly more refined version of this approach allows for the time in contact for the impact as this reduces the "flight" time. If the contact stiffness is k then half a cycle of contact vibration occurs in time $\pi\,(m/k)^{0.5}$ so the second equation becomes

$$2\pi/\omega - \pi\,(m/k)^{0.5} = 2\,V\,m/F$$

The biggest uncertainty occurs with the value of the coefficient of restitution at impact since effective masses are known. Once the impact velocity V and the contact (tooth) stiffness are known, the peak force can be estimated since by energy

$$0.5mV^2 = 0.5\,kx^2$$

where x is the maximum interference and the force is k x. For the first subharmonic response the flight time will correspond to two periods (i.e., 4 π/ω) less the contact time.

One danger with loss of contact is the possibility that the height of bounce is large enough to travel right across the backlash and impact on the unloaded trailing flank. A check on the meshing geometry of a standard spur gear pair shows that, as might have been predicted by the law of general cussedness, the impact on the trailing flank occurs at a time to inject a high return velocity and there is liable to be an extremely destructive hammer across the backlash. Fortunately this effect is extremely rare. Altering backlash may either improve matters or make the vibration worse.

It has been assumed in this description that the troublesome excitation is the classic 1/tooth but it is possible for a powerful phantom to have the same effect. Phantoms are produced when gear cutting machines have large once per tooth errors on their worm and wheel drives. Such phantoms are more likely to be troublesome on larger "industrial" gears and can produce subharmonics. Removal of phantoms is relatively straightforward but involves measuring the T.E. of the gear-cutting machine with portable T.E. equipment. Poor meshing profiles with an involute which is leant over can give a sudden lift in the T.E. curve which has the effect of throwing the gears out of contact due to the high upwards velocity associated with the sudden tip engagement.

11.4 Possible changes

The most obvious change is to reduce the T.E. if this is the cause of the trouble. This loss of contact depends on acceleration initially so it is desirable to compare the acceleration (torsional) due to any torsional vibrations (such as with a Diesel engine) with the acceleration due to the T.E., usually at 1/tooth but it could be due to harmonics or a phantom. Looking at the time pattern of the vibration trace will give a good idea of whether it is mainly 1/tooth repetitions or 1/rev or 2/rev that is causing the torsional acceleration which provokes the trouble.

If T.E. at 1/tooth is the cause, (it usually is) then measurement of T.E. will determine whether it is "reasonable" or excessive. The same considerations apply as in section 10.5 with economics controlling decisions. Changing spur gears to helicals or improving profile control may be possible but much depends on whether the existing T.E. is already good (< 5 μm ?) or poor.

Other parameters are often not directly controllable. The transmitted torque (and hence the force F trying to keep the teeth together) is determined by the load and so is not easily changed. The inertia of, say, a printing roll cannot be reduced. We are left with the problem that we cannot further reduce the acceleration due to the T.E. or, it seems, increase the F/m acceleration.

The two techniques occasionally possible are to increase F or reduce m. Increasing F, when the load is fixed, is possible only by recirculating power using the approach described in the next section, since using a brake would usually waste too much power. Decreasing m is not possible directly but may be possible by decoupling the large inertia of the driven load, or the motor from the gear by some form of elastic coupling. The necessary coupling must be very carefully designed since it must allow a high torsional natural frequency of the relatively light gear without allowing excessive lateral deflection of the gear or position inaccuracy of the driven load (the printing roll). This type of vibration decoupling design requires a high level of sophistication and is not always possible.

Occasionally it is possible to change tooth numbers to avoid trouble but this is less likely to be effective with non-linear systems than with linear systems and there is an inevitable stress penalty. Splitting a spur pinion and its mating wheel in two and staggering them half a circumferential pitch can sometimes reduce 1/tooth excitation. However, it is expensive and it is usually not possible to control eccentricity sufficiently, so changing to helical is usually more effective. Much depends on how good the helix alignments are as this is the major control factor with helicals.

11.5 Anti-backlash gears

The extreme case of low load can apply with control drives where the load may be zero for long periods. Any form of lost motion whether due to friction or backlash (or hysteresis) will make a servo control system very unhappy. The solution to prevent backlash in servos is the same as that to prevent non-linear bouncing oscillations in lightly loaded drives. In both cases the objective is to keep the gears firmly in contact without excessive wear rates.

The obvious solution is to make gears without backlash but this is not realistic. It is difficult to get the effective eccentricity of a mounted gear below 15 μm (0.6 mil) peak to peak, even with care and expense using reference shoulders, so with two gears the clearance can rise to 30 μm. Double flank interference contact must be avoided since wear and damage rates are then very high and bearings may also be damaged. Thermal effects are also significant since, with a temperature differential of 10°C on 200 mm centres, the extra growth would be 20 μm giving considerable extra loading on bearings and teeth if there is no initial clearance.

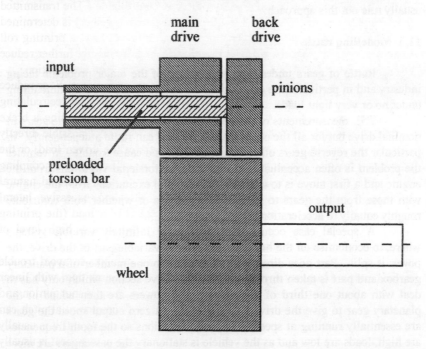

Fig 11.6 Sketch of torsion bar preloading of gear mesh to prevent loss of contact.

The technique that can be adopted is shown diagrammatically in Fig. 11.6. An additional gear is loaded with a torsion bar to impose sufficient load on the "back" face of the gear to keep the "working" face permanently in contact. In some designs the auxiliary gear is mounted on the main gear and sprung using a leaf spring design.

Extra support bearings and preloading the torque give difficulties for original design and for maintenance. Penalties are complexity, cost, bulk and a shortened lifecycle. On a bi-directional (servo) drive the "back" drive must be sprung with full working torque so the direct working gear has to be able to take twice full torque, and the gear system as a whole needs three times the torque rating of a single gear pair. Cycle life tends to be reduced because the back drive is operating under full load all the time, increasing wear and fatigue rates. On a normal unidirectional drive the back drive need usually not be as powerful but still has to operate all the time, decreasing gear life.

More complex systems can be devised using two servo drives in opposition but with programming control so that when drive is required in one direction the torque is removed from the other direction. Cost and complexity usually rule out this approach.

11.6 Modelling rattle

Rattle of gears under light load is one of the major problems facing industry and in particular the car industry since cars spend so much time idling under no or very light loads.

T.E. measurements of the gears are essential, not just for the gears in nominal drive but for all the other gears since they can rattle independently. In particular the reverse gears often have high T.E. and cause trouble. In vehicles the problem is often accentuated at idling by the torsional vibrations from the engine and a first move is to compare the torsional excitations from the engine with those from the gears to see which dominates or whether both contribute roughly equally to accelerations.

A special case occurs with split drive infinitely variable systems where, to economise on the heavy and expensive variable part of the drive, the power is split. Part goes directly through gears to one member of a planetary gearbox and part is taken through the variable drive section which only has to deal with about one third of the power. The powers are then added in the planetary gear to give the drive to the wheels. At zero output speed the gears are essentially running at speed in opposite directions so the tooth frequencies are high, loads are low and as the vehicle is stationary the passengers are more likely to be aware of any noise.

As the problem is non-linear and complex there is a requirement to model the system so that the effects of changes can be estimated at least roughly without the delays and costs of cutting metal each time. This is more complicated than it sounds as in the standard transverse engined car there are two meshes in drive and several others running free. Modelling the complete system would involve considering both torsional and lateral movements with allowance for 3-dimensional effects and so would be extremely complex. Such systems exist [1] but are very complex and time consuming to program and hence expensive so can only be used economically for mass production requirements.

Investigations of problems can be much simplified by reducing the model to one in which there are only torsional movements of the gears possible. This is reasonable for the final drive of a transverse engined car but is less representative for the intermediate gears which are on shafts which flex significantly laterally.

The resulting simplest possible model is shown in Fig. 11.7. This assumes rigid bearings (with no play), that input from the engine can be modelled as a torque Q with an input moment of inertia 1 and that at output the wheels are effectively fixed so that the differential crown-wheel (5) is connected to "earth" via the torsional flexibility of the drive shafts.

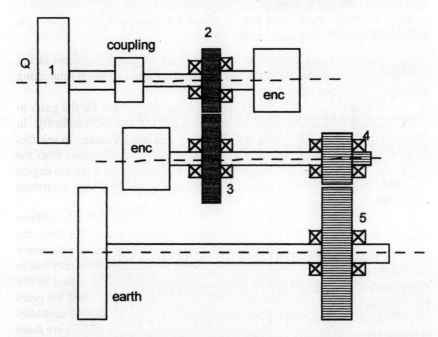

Fig 11.7 Simplest model of transverse engine drive system with two non-linear meshes and torsional oscillations at input.

The model should allow for the insertion of a T.E. at meshes 2 - 3 and 4 - 5 and to model the effects of the main engine torsionals a Hooke's coupling will give 2/rev excitation if misaligned. Unfortunately this does not duplicate the rapid changes associated with firing. In the laboratory there is easy access to shaft ends so encoders can be fitted as shown in the diagram and an encoder can also be fitted to the output shaft at the crown-wheel 5. Getting instrumentation on a real engine is relatively easy at positions 2, 3, and 4 but is almost impossible at position 5. The choice between encoders and tangential accelerometers is difficult for this type of rig as encoders are better for the initial determination of quasi-static T.E. but for detecting sudden accelerations and impacts, accelerometers are preferable.

The corresponding equations are of the form:

All measured clockwise. r is base circle radius, I inertia, k angular stiffness, K contact stiffness, D is angular damping coefficient, A is angular acceleration, V is angular velocity, s is angular displacement. F is contact force at a mesh.

Single suffix to earth, double is relative. te12 is TE due to coupling
te23 and te45 are due to meshes

Input Q, inertia 1, shaft, input gear 2, lay gear 3, shaft, differential pinion 4, differential wheel 5, half shaft, earth.

Motion

I1 A1 = Q - D1 V1 -k12 (s1-s2 +te12) - D12 (V1-V2)   rearranges to
I1 A1 + D1 V1 + k12 (s1-s2+te12) + D12 (V1-V2) = Q   and similarly
I2 A2 + D2 V2 - k12 (s1-s2+te12) - D12 (V1-V2) = - F23 r2
I3 A3 + D3 V3 + k34 (s3-s4) + D34 (V3-V4) = - F23 r3
I4 A4 + D4 V4 - k34 (s3-s4) - D34 (V3-V4) =   F45 r4
I5 A5 + D5 V5 + k5 (s5) + =   F45 r5

Divide throughout by base circle radii to get "linear" equations and take r1=r2

$[I1/r2^2]$ (A1.r2) + $[D1/r2^2]$ (V1 r2) + $[D12/r2^2]$ (V1r2-V2r2) + $[k12/r2^2]$ (s1r2-s2r2+te)   = Q/r2
$[I2/r2^2]$ (A2.r2) + $[D2/r2^2]$ (V2 r2) + $[D12/r2^2]$ (V2r2-V1r2) + $[k12/r2^2]$ (s2r2-s1r2-te)   = - F23
$[I3/r3^2]$ (A3.r3) + $[D3/r3^2]$ (V3 r3) + $[D34/r3^2]$ (V3r3-V4r3) + $[k34/r3^2]$ (s3r3-s4r3)   = - F23
$[I4/r4^2]$ (A4.r4) + $[D4/r4^2]$ (V4 r4) + $[D34/r4^2]$ (V4r4-V3r4) + $[k34/r4^2]$ (s4r4-s3r4)   = + F45

$$[I5/r5^2] (A5.r5) + [D5/r5^2] (V5 \; r5) +$$
$$[k5/r5^2] (s5r5) \qquad\qquad = + F45$$

$$[M] \quad [A] = -[Dabs] [V] \quad - [Drel][V] + [Drel][Vtr] - [Krel][X] +$$
$$[Krel][Xtr] \qquad\qquad = [F]$$

Tooth forces

$$F23 = K23 \; [s2 \; r2 + s3 \; r3 + te23] + D23 \; [V2 \; r2 + V3 \; r3]$$
$$F45 = - K45 \; [s4 \; r4 + s5 \; r5 + te45] - D45 \; [V4 \; r4 + V5 \; r5]$$
If negative, force is put to zero.

Combined

$$A = [\; F - Dabs.*V - Drel.*V + Drel.*Vtr - Krel.*X + Krel.*Xtr] /[M]$$
$$V = V + tint*A; \qquad\qquad X = X + tint*V.$$

The equations above can be programmed by the standard time marching approach as in chapter 5 to give dynamic responses to the assumed errors. The same problems arise in that the starting positions and velocities chosen will give long settling times unless initial torsional windups are considered but as these are small with the light mean loads involved in rattle the settling is faster. As discussed previously the dominating problem is to set realistic damping levels. With high speed impacts the system in practice no longer behaves as lumped masses and springs. The impacts tend to radiate energy in the form of shock waves where little energy returns to the shock source so the apparent damping is high.

A typical program is

```
% NON-LINEAR VERSION
% Rat4  Rattle equations, added damping, all angles clockwise, backlash
% inertia-1, shaft, input gear2, layshaft gear3, shaft, diff pinion4,
% diff wheel5, half shaft, earth. Setup parameters  2micron TE
clear;    % equivalent linear masses
M = [ 6.3 0.63 1.0 1.2 5 ] ;        % pi*0.045(4th)*0.02*7840/(2*0.04sq) kg
Dabs = [ 200 100 100 100 100] ;    % start low damping freq order 30 Hz
Drel = [300 300 300 300 0] ;% rel shaft damping,1-2 3-4 freq order 400,30 Hz
K = [8e6 8e6 2e6 4.5e6 1e6 ];       % shaft stiffnesses 1-2,3-4,5-earth/r(sq)
% turned into equiv linear stiffnesses at teeth
% T/lrbsq = 81e9*pi*0.01(4th)/2*0.1x0.04(sq) for 1-2  torsional
tint = 5e-5; % time step interval. max before instability?
CF = 40 ;  % input contact force equivalent Q/rb
bl1 = 3e-5 ; bl2 = 4e-5 ;            % 30 micron backlash
rev = input('Input revs/sec    ');    % Angle is rev x teeth/rev x 2pi x time
% set input rev to rev/s then tors is 2*rev*2*pi rad/s
```

```
% 1st tooth is 29*rev*2*pi rad/s  2nd is 17*rev*2*pi rad/s
tors=12.6*rev*tint ;   tooth1= 182*rev*tint;      tooth2= 107*rev*tint;
A =[0 0 0 0 0];V=[0 0 0 0 0];X=[3.1e-4 2.9e-4 -2.9e-4 -1.2e-4 1.2e-4];% initial
Z = round(8/(rev*tint)) ;         % number of points in sequence for 8 rev
seq = zeros(5,Z);   force = zeros(2,Z);          % setup final results
for n = 1:Z;                      % +++++++++++++++++++ start time step loop
te12=5e-5*sin(tors*n); % due to 2/rev torsionals ~ 100 micron.
te23=2e-6*sin(tooth1*n);         % TE 4 μm p-p
te23r=2e-6*sin(tooth1*n + 3);% reverse about πι lag
te45=2e-6*sin(tooth2*n);te45r=2e-6*sin(tooth2*n + 3);%TE +ve for +ve metal
Xtr = [(X(2)-te12) (X(1)+te12) 1.5*X(4) 0.67*X(3) 0];  % includes coupling
Vtr = [V(2) V(1) 1.5*V(4) 0.7*V(3) 0];
if X(2)+X(3)+te23 > 0;                           % drive flank +ve force
       F23 =  2e8*(X(2)+X(3)+te23) + 3e2*(V(2)+V(3));
       elseif X(2)+X(3)-te23r+bl1 < 0;           % overrun flank -ve force
   F23 = 2e8*(X(2)+X(3)-te23r+bl1) + 3e2*(V(2)+V(3));
   else
   F23 = 0;                                      % in backlash
end
if X(4)+X(5)+te45 < 0;                           % drive flank
   F45 = -3e8*(X(4)+X(5)+te45) - 3e2*(V(4)+V(5));
       elseif X(4)+X(5)+te45r - bl2 > 0;         % overrun flank
   F45 = -3e8*(X(4)+X(5)+te45r-bl2) - 3e2*(V(4)+V(5));
   else
   F45 = 0;                                      % in backlash
end
F = [CF -F23 -F23 F45 F45];     % ext and tooth forces
A = (F - Dabs.*V - Drel.*V + Drel.*Vtr -K.*X + K.*Xtr)./M;  % acelerations
V = V + tint*A ;        X = X + tint*V;
seq(:,n) = (X');  % stores displacements for plot
force(1,n) = F23 ;      force(2,n) = F45 ;      % mesh forces
end                             % +++++++++++++++++++++++++ end time step loop
ser = 1e6*(seq') ;xx = (1:n)*tint*1000;  % x axis in millisec
last = round(ser(Z,:))            % displ starting conditions for next try
figure;  plot(xx,ser);
xlabel('time in milliseconds'); ylabel('displacement in microns'); pause
figure;   plot(xx,force);
xlabel('time in milliseconds'); ylabel('tooth force in Newtons'); pause
single = round(Z/8); begin = Z - single;
xx1 = xx(begin:Z); ser1 = ser((begin:Z),:);force1 = force(:,(begin:Z));
figure;   plot(xx1,ser1);
xlabel('time in milliseconds'); ylabel('displacement in microns'); pause
```

figure;   plot(xxl,force1);
xlabel('time in milliseconds'); ylabel('tooth force in Newtons');
avgF = sum (force(1,(1:Z)))/Z   % checks mean force right
% colours 1 - blue, 2-green, 3-red, 4-turqoise, 5-purple.

The results from such a program are shown in Fig. 11.8 for a rather extreme case of inaccurate gears at high speed under a low mean contact load in the first mesh of 40 N (9lbf) where the gears are hammering across the backlash zone so there are negative tooth forces. As expected peak magnitudes are far above the mean levels.

Modelling such systems is not difficult and there have been many models but what is lacking is experimental verification so any model should be treated with great caution.   Uncertainties about lateral deflections, any 3-D axial effects and complete ignorance of effective damping in the impacts do not assist reliability.

Unlike the estimates of chapter 5 there has been no attempt to model the fine details of the mesh contacts because the impacts are extremely short and high force so the contact will be right across the full facewidth and so a constant stiffness assumption is reasonable.

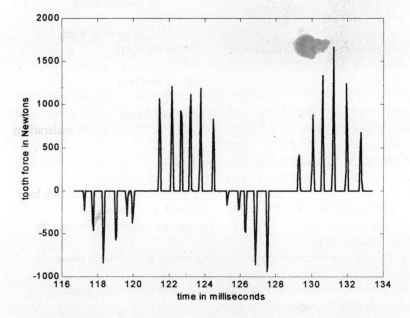

Fig 11.8 First mesh tooth forces at 3600 rpm as modelled on computer.

Reference

1.   Romax Ltd., 67 Newgate, Newark NG24 1HD.   www.romaxtech.com

12

Planetary and Split Drives

12.1 Design philosophies

The conventional parallel shaft gear drive works well for most purposes and is easily the most economical method of reducing speeds and increasing torques (or vice versa). The approach starts running into problems when size and weight are critical or when wheels start to become too large for easy manufacture. If we take the torques of the order of 1 MN m (750000 lbf ft) that are needed for 6000 kW (8000 HP) at 60 rpm we can estimate the wheel size for a 5 to 1 final reduction. The standard rule of thumb allows us about 100 N mm^{-1} per mm module so assuming 20 mm module (1.25 DP) gives us a wheel face width of about 450 mm and diameter of 2.25 m. This is not a problem but if the torque increases we rapidly reach the point where sizes are too large for manufacture and satisfactory heat treatment especially as the carburised case required thickness also increases.

The solution is to split the power between two pinions so that loadings per unit facewidth remain the same but the torque is doubled. The further stage in this approach is to split the power between four pinions to give roughly quadruple increase in torque without significant increase in size. This fits in well if there is a double turbine power drive which is often wanted for reliability. The design is as sketched in Fig. 12.1. Power comes in via the two pinions labelled IP, splits four ways to the four intermediate wheels (IW) which in turn drive the four final pinions which mesh with the final bull wheel. The resulting design is accessible and reasonably compact though at the expense of extra complexity in shafts and bearings.

To achieve the gains desired with power splitting it is absolutely essential that equal power flows through each mesh in parallel so as there are inevitable manufacturing tolerances, eccentricities and casing distortions some form of load sharing is needed. This is usually conveniently and easily provided by having the drive shafts between intermediate wheels and final pinions acting as relatively soft torsional springs. If the accumulated position errors at a mesh sum to 100 μm and we do not want the load on a given pinion to vary more than 10% the torsional shaft flexibility must allow at least 1 mm flexure under load.

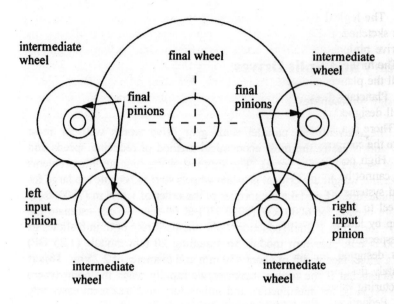

Fig 12.1 Multiple path high power drive.

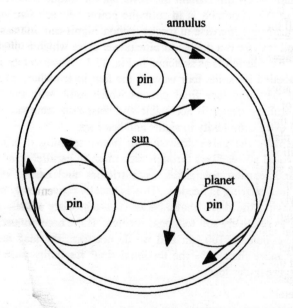

Fig 12.2 Typical planetary drive showing forces on planets.

The logical extension of the multiple path principle is the planetary gear as sketched in Fig. 12.2 where to reduce size (and weight) further the final drive pinions are moved inside the wheel which becomes an annular gear. The further asset of the planetary approach is that a single sun gear can drive all the planets and with 3 planets the reduction ratio can be as high as 10 : 1. Planetary designs give the most compact and lightest possible drives and well designed ones can be a tenth the size and weight of a conventional drive. There is a corresponding penalty in terms of complexity and restricted access to the components.

High performance is again dependant on having equal load sharing but this cannot be achieved by torsion bar drives and so there are many "best" patented systems for introducing load sharing. The simplest is to allow the sun wheel to float freely in space so that any variations in meshing can be taken up by lateral movements of the sun. More commonly in high power drives especially as designed by Stoeklicht, the annulus, which is relatively thin, is designed to flex to accommodate variations. A third variant deliberately designs the planet supporting pins to be flexible to absorb any manufacturing variations.

Pedantically the term "planetary gear" is used to describe all such gears whereas the more commonly used "epicyclic" is only correct for a stationary annulus and if the planet carrier is stationary it is a star gear. When a gear is used in an infinitely variable drive as a method of adding speeds then all three, sun, annulus and planet carrier are rotating.

12.2 Advantages and disadvantages

The advantages of splitting the power are mentioned above in terms of reduction of weight and size and frontal area (for aeroplanes and water turbines) and the corresponding disadvantages of increased complexity and, in the case of planetary gears, poor accessibility.

Additional factors can be the problems of bearing capabilities since as designs are scaled up the mesh forces and hence the bearing loads tend to rise proportional to size squared whereas the capacity of rolling bearings goes up more slowly and the permitted speeds decrease. This imposes a double crimp on design and forces designers towards the use of plain bearings with their additional complications. Splitting power delays the changeover from rolling bearings to plain bearings for the pinions and as the pinions can be spread around the wheel the wheel bearing loads can be reduced or in the case of planetary gears the loads from the planets balance for annulus and planet carrier completely.

The planet gears are very inaccessible and are highly loaded so they present the most difficult problems in cooling. For high power gears it is normal to have the planet carrier stationary as this makes introducing the large quantities of cooling oil required much easier.

There would appear to be no obvious limit to power splitting but in an external drive it is complex to arrange to have more than four pinions and even this requires two input drives. Planetary gears can have more than three planets and five are occasionally used. However load sharing is still needed and, as the system is redundant, cannot be achieved by floating the sun so either the planet pins must be flexible or the annulus must flex. There is the additional restriction that with five planets the maximum reduction (or speed increasing) ratio is limited by the geometry to slightly less than five if used as a star gear or five if an epicyclic. Design problems can arise with heavily loaded planets because with most designs it is necessary to support the outboard ends of the planet pins and the space available between the planets for support structure is very limited as can be seen in Fig. 12.3.

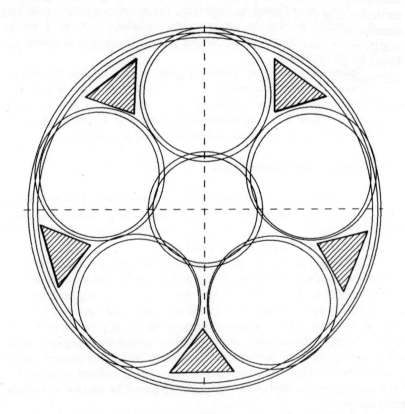

Fig 12.3 Maximum reduction with five planets.

Care must also be taken that the planet carrier is rigid so that the outboard support members are not allowed to pivot at their base when under load.

Planetary gears automatically have input and output coaxial which can be either an advantage or disadvantage according to the installation. The fact that the reaction at the fixed member, whether annulus or carrier, is purely torsional can be a great advantage for vibration isolation purposes as a very soft torsional restraint can be used to give good isolation without fear of misalignment problems.

12.3 Excitation phasing

If we have three, four or five meshes running in parallel there will be the corresponding number of T.E. excitations forcing the gear system and attempting to produce vibrations to cause trouble. It is easiest to consider a particular case such as the common three planet star drive and to make the assumption that the design is conventional with the three planets spaced exactly equally and that spur gears are used. If we then look at the vibrating forces on the sun we have the three forces as shown in Fig. 12.4, spaced at 120° round the sun and inclined at the pressure angle to the tangents.

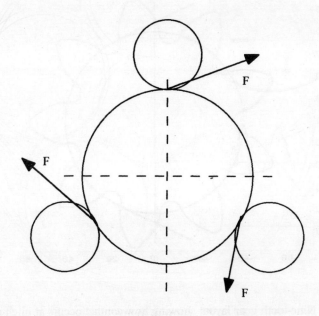

Fig 12.4 Sun to planet force directions.

The three meshes will probably have roughly the same levels of T.E. and so the same vibration excitation and will have the same phasing of the vibration relative to each pitch contact. The three pitch contacts can be phased differently according to the number of teeth on the sunwheel. If the number of teeth on the sun is divisible by three the three meshes will contact at the pitch point simultaneously and the three excitations will be in phase. This will give a strong torsional excitation to the sun but no net sideways forcing.

If not, the three excitations will be phased 120° of tooth frequency apart in time and at 120° in direction so there will be no net torsional vibration excitation on the sunwheel but a vibrating force which is constant in amplitude and whose direction rotates at tooth frequency. The direction of rotation is controlled by whether the number of teeth is 1 more or 1 less than exactly divisible by 3.

The same considerations apply for the three mesh contacts between the planets and the annulus. Dependent on whether the number of annulus teeth is exactly divisible by three or not we can choose to have predominantly torsional vibration or a rotating lateral vibration excitation.

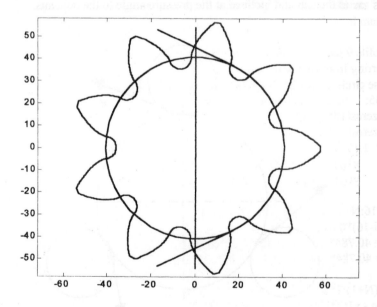

Fig 12.5 Nine-tooth gear layout showing how contact occurs at pitch points at roughly the same time.

When there are five planets there are similar choices as to whether the excitations are phased or not to give predominantly torsional vibration or lateral vibration. The choice should depend on whether the installation is more sensitive to torsional or lateral problems.

Similar considerations apply for the planets where the 2 meshing excitations on a planet can either be chosen to be in phase or out of phase. The former gives tangential forcing on the planet support, the carrier, while the latter gives rotational forcing on the planet itself which being light can usually rotate easily. As the contact is on the opposite flank it is not immediately obvious whether an odd or even number of teeth is needed on the planet but an odd number of teeth will give simultaneous pitch point contact to sun and annulus and an even number will give 180° phasing and so less torsional excitation on the carrier. Fig. 12.5 shows the rather extreme case of a nine-tooth 25° pressure angle gear which is meshing on both sides as in a double rack drive or as in a planet (though it would not be normal to use less than about eighteen teeth in practice).

The pressure lines are shown tangential to the base circle and it can be seen that contact (along the pressure lines) will occur at the (high) pitch points at roughly the same instant in time so there will be low net tangential forces on the planet but sideways forcing on the planet pin. The Matlab program to lay out the pinion is

```
% profile 9 tooth  10 mm module  25 deg press angle
% starting from root with radius 5
% base circle 45 cos 25  = 40.784  root centre -5, 40.784
N = 65;    % no of points for each flank.
x1 = zeros(18*N,1);
y1 = zeros(18*N,1);
for i = 1:15                              % root circle
        x1(i) = -5 + 5*cos(1.4488 -(i-1)*0.1);
        y1(i) = 40.784 - 5*sin(1.4488 -(i-1)*0.1);
end
for i=16:N ;                  % involute
ra = (i-16)*0.02 ;
x1(i)= 40.784*(sin(ra)-(i-16)*0.02*cos(ra));
y1(i)= 40.784*(cos(ra)+(i-16)*0.02*sin(ra));
end
for i=(N+1):2*N ;                         % Image in x=0 other flank
x2(i) = - x1(2*N+1-i);   y2(i) = y1(2*N+1-i);
rot1 = 0.45413 ;
x1(i) =x2(i)*cos(rot1) +y2(i)*sin(rot1);
y1(i) = -x2(i)*sin(rot1) +y2(i)*cos(rot1);
```

```
end
for th =1:8;                              % rotate for other 8 teeth
x1((th*2*N +1):(th+1)*2*N) =
x1(1:2*N)*cos(0.69813*th)+y1(1:2*N)*sin(0.69813*th);
y1((th*2*N +1):(th+1)*2*N) =-
x1(1:2*N)*sin(0.69813*th)+y1(1:2*N)*cos(0.69813*th);
end
save teeth9 x1 y1
for ang = 1:44                            % plot base circle
xo(ang) = 40.784*cos(ang*0.15); yo(ang) = 40.784*sin(ang*0.15);
end
xt1 = [17.236 -17.236]; yt1 = [36.963 53.037] ;      % tangent
xt2 = [17.236 -17.236]; yt2 = [-36.963 -53.037] ;    % tangent
ax1 = [0 0] ; ax2 = [-54 54]; %      vertical axis
phi = -0.05 ;                          % rotate gear to symmetrical position
u2 = x1*cos(phi)+y1*sin(phi) ; v2 = -x1*sin(phi)+y1*cos(phi);
figure
plot(u2,v2,'-k',xo,yo,'-k',xt1,yt1,'-k',xt2,yt2,'-k',ax1,ax2,'-k')
axis([-58 58 -58 58])
axis('equal')
```

12.4 Excitation frequencies

For simple parallel shaft gears it is easy to see what the meshing frequencies will be as they are rotational speed times the number of teeth. In a planetary gear there will be at least two and possibly three out of the sun, planet carrier and annulus rotating so the tooth meshing frequency is less obvious.

The simplest case occurs with a star gear as the planets, though rotating are stationary in space. In Fig. 12.6 with S sun teeth, P planet teeth and A annulus teeth, the ratio will be A/S and as 1 rotation of the sun will involve S teeth, the frequency will be S times n where n is the input speed in rev s^{-1}. This is the same as A times R where R is the output speed which will be in the opposite direction but this does not alter the meshing frequency.

When the planet carrier is rotating then both the sun to planet mesh and the planet to annulus mesh are moving in space so there is not a simple relationship and we must first bring the carrier to rest. As before, with the carrier at rest the tooth frequency will be S times n where n is the input (sun) speed relative to the (stationary) carrier. On top of this we impose a whole body rotation to bring the carrier up to the actual speed and the other gears will also have this speed added but the meshing frequency will not be altered as it is controlled solely by the relative sun to carrier speed.

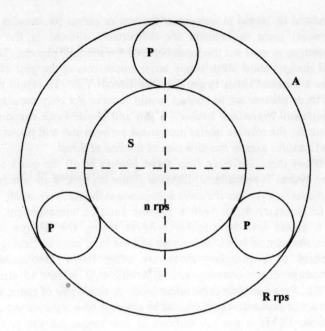

Fig 12.6 Sketch of planetary gear for meshing frequencies.

The general relation between speeds is determined relative to a 'stationary' carrier. Then with speeds ω

$$(\omega_s - \omega_c) / (\omega_a - \omega_c) = -R \qquad \text{where } R = N_a / N_s$$

or

$$\omega_s = (1 + R) \omega_c - R \omega_a$$

In general, whatever the speed we take the (algebraic) difference between sun and carrier speeds and multiply by the number of teeth on the sun to get the tooth meshing frequency or the corresponding difference between carrier and annulus speeds and multiply by the number of teeth on the annulus.

12.5 T.E. testing

Complications arise if the T.E. of a complete planetary or split drive is required because there are several drive paths in parallel under load.

If the drive is as sketched in Fig. 12.6 and there is an error in one of the three sun-to-planet meshes, we will not necessarily detect a relative torsional movement between sun and planet.   The error may be accommodated by lateral movement of the sun or planet (or annulus flexing or movement) since movements are deliberately allowed in the various (patent) designs to even out the loads between the multiple planets. The most successful designs allow surprisingly large movements of the gear elements, sometimes a hundred times larger than the 1/tooth T.E. To obtain the T.E. when all three planets are in contact would involve not only measuring the relative torsional movement between a sun and planet (with encoders), but also measuring the relative lateral movement between sun and planet axes or planet and annulus axes in the direction of the line of thrust.

When there are more than three planets or all the gears are held rigidly the system is redundant. Either a planet support or an annulus must flex or a planet lose contact if elastic deflections at the teeth are small.

In planetary drives with a flexible annulus, measurement of T.E. between a planet and the annulus involves taking the relative torsional movement, the relative lateral movement and the local annulus flexing. Since the members of a planetary drive are often rather inaccessible, this instrumentation is too complex and difficult, so it is rare to attempt to measure T.E. for a complete drive under load. A single pair of gears, whether sun-to-planet or planet-to-annulus must be checked on a separate test rig with fixed centres. This is not too difficult at low torque but the problem of driving a large planet at full torque against an annulus while maintaining alignment and positions yet leaving access for encoders is almost impossible. Planets on large drives do not normally have provision for transmitting torque as the loads on a planet are balanced and driving torque from one end is likely to give spurious results due to planet windup which does not occur in position. When the planet to annulus mesh is loaded there is the additional factor of (design) annulus distortion to complicate life.

Split drives present similar problems though access is usually much easier and axes are held rigidly so that there are not the complications of lateral movements but unless the pinion drive torque shafts are flexible there is the possibility of uneven load distribution between the pinions. Similar considerations apply for testing double helical gears as they are effectively two gears working in parallel and for anti-backlash sprung drives the sections of the gear must be tested separately if the combined unit shows errors.

12.6 Unexpected frequencies

With any gear drive we normally expect to encounter noise trouble from tooth frequency and its harmonics with modulation sometimes giving sidebands spaced 1/rev either side of the tooth frequency harmonic. There may also be phantom or ghost frequencies present due to manufacturing imperfections or occasionally in high speed gears there may be pitch effects (see section 9.10). All these will normally be picked up easily by conventional T.E. testing which can be under low load as these effects are not normally altered by loading.

In the case of planetary gears there is also the possibility of amplitude modulation due to the passing frequency of the planets. We can consider an epicyclic gear as in Fig. 12.7 with five planets and an accelerometer detecting vibration on the stationary (moderately flexible) annulus or a connection transmitting vibration to the rest of the installation at one position on the annulus.

The vibration observed will be highest when the excitation from a planet is near the accelerometer and will reduce between planets so there will be an amplitude modulation of the signal at 5 times per revolution of the carrier. This will appear as sidebands spaced either side of the tooth frequency and should be relatively easy to identify.

There is a rather more subtle effect that can occur due to the variable position of the accelerometer relative to the excitation from the planets. This effect can be explained in the frequency domain by analysing the effect of the excitation from the mesh being multiplied by the time varying transmission path between the mesh and the accelerometer. The theory is given by McFadden in Ref. [1].

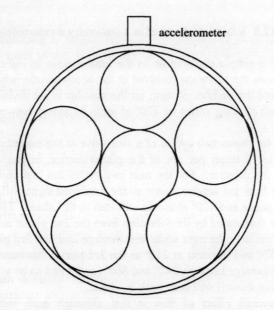

Fig 12.7 Diagram of five planet epicyclic.

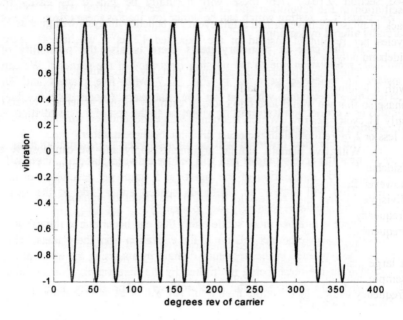

Fig 12.8 Vibration observed at a stationary accelerometer.

There is simple explanation in the time domain as indicated in Fig. 12.8 which shows the vibration received at the accelerometer when there are three planets and the number of teeth on the annulus is not divisible by three so that the mesh phasing varies by 120° of tooth frequency between the three planets.

The plot shows two cycles of a sine wave at the expected frequency, in this example 12 times per rev of the planet carrier, in phase as the 1st planet is near the detector. For the next two cycles the first and 2nd planet are equidistant from the accelerometer so the combined signal will have equal amounts of in-phase and 120° phase so will sum to 60° phase. The next sixth of a rev will be dominated by the vibration from the 2nd planet and so will be 120° phase. Similarly the next sixth will average 2nd and 3rd planet phases and so be at 180° and the next at 240° as the 3rd planet dominates, while the final sixth will average between 240° and 360° (or 0°) and so be at 300° phase. The next rev (not shown) will start back in-phase.

The overall effect of this is that although each section of the vibration is oscillating at exactly twelve per rev of the carrier, the phase changes (technically a phase modulation) give a different frequency. Counting up the cycles shows that there are 12 and two-thirds cycles of oscillation in the revolution and as the next sector vibration will be exactly back in phase we have gained a full cycle. Frequency analysis will give us 13 cycles per revolution instead of the expected 12 as if we had an upper sideband only.

The above description assumed that the succeeding planets came with a leading phase but equally well the planets could come with a lagging phase so that we would lose one cycle in each rev of the carrier and observe only 11 cycles per rev. Which frequency we get depends on whether there is 1 less or 1 more tooth than the exact divisible by three number.

With five planets, similar arguments apply and we can observe a "sideband" with 1 more or 1 less cycle than the "correct" value. There is however the possibility of having two more or two less teeth than exactly divisible by five and we would then get the result of apparently a single frequency at two cycles more or less per carrier rev than the expected frequency.

The frequency obtained when carrying out a frequency analysis with a large gear will depend on the length of time of the sample since a short sample may effectively be from one planet only and so may be at tooth frequency while a long sample from a full rev will be as described above

Reference

1.   McFadden, P.D. and Smith, J.D., 'An Explanation for the Asymmetry of the Modulation Sidebands about Tooth Meshing Frequency in Epicyclic Gear Vibration', Proc. Inst. Mech. Eng., 1985, Vol. 199, No. C1, pp 65-70.

13

High Contact Ratio Gears

13.1 Reasons for interest

We normally define the "geometrical" contact ratio between a pair of gears as the length of line of contact, measured along the pressure line, from pinion tip to wheel tip, divided by the distance between two successive teeth surfaces also measured along the pressure line, i.e., the base pitch.

From a glance at Figs. 2.7 and 2.8 it is obvious that contact normally does not go anywhere near the tip of either tooth and that real contact ratios are much lower than nominal contact ratios. At low loads in particular, there is only one pair of teeth in contact. More typically, under load, a nominal contact ratio of 1.7 might give double contact for only 15% of the time instead of the expected 40% (0.7/1.7).

With conventional proportion teeth neither "short" nor "long" relief can give low T.E. at both high and low load, but if we can get the true contact ratio up to about 2.0, then it is possible to have quiet running at high and low loads. It is not possible to get a nominal contact ratio above 2.0 (and hence a true contact ratio of about 2) because the original standard tooth proportions and pressure angle were chosen rather arbitrarily a century ago well before gear meshing was investigated and an understanding gained [1].

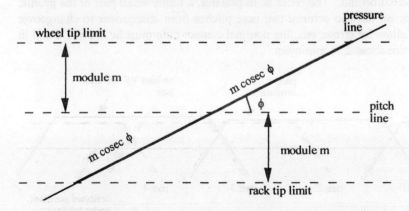

Fig 13.1 Length of contact line with large tooth numbers or rack.

Fig. 13.1 shows the limiting case for standard teeth when a very large gear mates with another large gear or a rack. A pressure angle of 20° and an addendum equal to the module gives the maximum length of approach and recess as m cosec ϕ and as the base pitch is m π cos ϕ the limiting ratio is 0.99 so the contact ratio cannot exceed 1.98.

The idea behind high contact ratio gears is that for high (design) loads we can apply the standard Harris map approach as in Chapter 2 and design the tip reliefs so that the elastic deflections at changeover are compensated by the shape of the relief. At low loads the contact cannot "drop" into the shape left by the tip reliefs as there is a third pair of teeth in the middle of their contact roughly at their pitch point and so maintaining the contact on the pure involute. At the changeover under load there are two pairs of teeth each taking half of the design force and the intermediate pair of teeth is taking the full design force, which is half of the total contact load directed along the pressure line.

We thus have the possibility of very low T.E. at design load and at very low load with a relatively low T.E. at intermediate loads compared with standard spur gears.

13.2 Design with Harris maps

Fig. 13.2 demonstrates the principle; successive teeth have been staggered slightly (a pitch error) for clarity. At low load there is always one mating pair of teeth on the "pure involute" so there is zero T.E. At high load there is "long relief" to give a smooth changeover from one pair to the pair two teeth behind. The relief is, in practice, a rather small part of the profile. Since we have to achieve two base pitches from changeover to changeover and allow for errors, etc., the nominal contact ratio must be above 2.0 and in practice about 2.25 minimum.

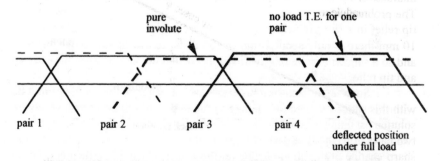

Fig 13.2 Geometry of tip relief and deflections for contact ratio of 2.

As with standard gears the amount of tip relief at the tips must allow for elastic deflections under maximum loads, pitch errors, profile errors and increased deflections due to overloads or misalignments. The amount of relief at the changeover points should be governed solely by the average expected elastic deflection when there are two pairs of teeth in contact (away from the changeover) and should be half this value.

This type of spur gear design will give low noise under a range of loads and is reasonably insensitive to alignment errors though it requires accurate manufacture.

The ideas behind designing very quiet spur teeth by achieving a real contact ratio of 2.0 have been understood in principle since the detailed dynamic work (by Gregory, Harris and Munro) was published more than 40 years ago, and in industry work was done as long ago as 1949 at Wright Aero [see chronology in Ref. 2]. Some 20 years later Boeing pursued the concepts, but it was only recently that Munro and his student Yildirim at Huddersfield [3,4] succeeded in measuring the actual unloaded and loaded T.E., quasi-statically and dynamically for extremely accurately manufactured gears, together with the corresponding vibration levels, to show that they are exceptionally quiet even by modern standards.

13.3 Two-stage relief

Another slight modification to the philosophy involving high contact ratio and a two-stage tip relief can, in theory, give zero T.E. at not only full load but also at an intermediate load. Work by Munro and Yildirim investigated this possibility.

There is a corresponding disadvantage in that there is some T.E. at zero load and there is a stressing penalty involved.

Another variant uses a two-stage relief to ease one of the manufacturing problems that can occur with the design shown in Fig. 13.2. The problem arises because the tip relief design has to give perhaps 50 μm of tip relief in a short roll distance. If the base pitch of the gear is of the order of 10 mm there is only about 1 mm of roll distance in which to move the 50 μm and so a sudden change of direction is needed at the join between pure involute and tip relief.

Some manufacturing processes which generate the profile cannot deal with this sudden a direction change so the design has to be altered. A possible solution is indicated in Fig. 13.3 which shows the changeover area between two pairs of teeth. The dashed lines are the basic tip relief shape and involve a sharp change of direction of relief at the points labelled E, which are the ends of the pure involute sections.

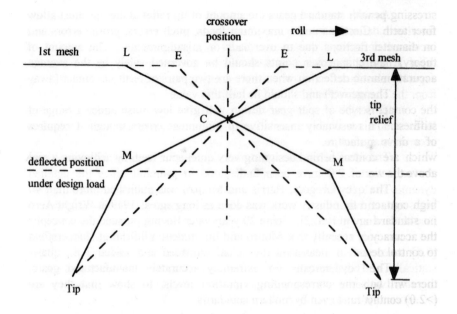

Fig 13.3 Sketch of basic tip relief design and modified shape.

As previously, there is no T.E. at light load because the intermediate pair of meshing teeth are in the middle of their involute section. It is not possible to alter the amounts of the tip reliefs or to alter the deflection at the crossover point C so these points remain fixed.

The modification to ease manufacturing is to start the tip reliefs at the points labelled L, roughly 1.5 times as far as the E points from the crossover. The tip relief then increases at about two-thirds the previous rate until the M points are reached at full depth of the expected deflected position. The tip relief then continues to the fixed tip position. The angle change (formerly at E) is reduced to two-thirds of previous and there is a comparable angle change at M which is also two-thirds of the previous E angle. Although two angle changes are now involved, the second one at M is in a less critical part of the relief and so errors of shape are unimportant.

13.4 Comparisons

To get 2.2+ nominal contact ratio we need taller, thinner teeth and the pressure angle must come down to below 16° with large numbers of teeth, or the teeth must be taller, again pushing up the minimum number of teeth to prevent pointed teeth. This means in general more teeth, which involves using a lower module. There is thus a double root stressing penalty as there is a stressing penalty associated with slender teeth and the penalty associated with finer teeth. Contact stresses are relatively unaffected as they depend primarily on diameter rather than module. There is a stressing bonus from always (in theory) having two pairs of teeth in contact but this depends on having accurate manufacture and very low adjacent pitch errors.

The other factor which is affected by the use of high contact ratios is the contact flexibility of the mesh as tall slender teeth greatly reduce the mesh stiffness. This will have a negligible effect on the lower resonance frequencies of a drive system but will reduce the frequencies of those vibration modes which are controlled by tooth stiffness. These frequencies are usually well above the working range for most gears.

The question is sometimes asked as to whether it is better to go to high contact ratio (spur) gears or to helicals if noise is very critical. There is no standard answer because the main factor controlling helical gear noise is the accuracy of alignment, assuming well designed gears. This is very difficult to control despite its dominating effect on both noise and stresses.

The "best" answer in a critical case is to be pessimistic and assume there will be some alignment errors and to make the drive helical with a high (>2.0) contact ratio. As always, design is a trade-off between noise and stress.

13.5 Measurement of T.E.

For conventional gears measurement of T.E. in the metrology lab. is straightforward as we mount them at the correct centre distance. Although we only see the zero load T.E. as in Fig. 2.7 the shape tells us the important information which is how large the T.E. dip is at the changeover point and whether the tip relief has started the correct roll distance from the pitch point. If these are correct we can reasonably infer that the part of the involutes we cannot see (dashed) is probably good enough.

High contact ratio gears immediately present a problem since if we mount them at the correct design centre distance then under no load we should get zero T.E. right through the meshing cycle so we cannot see if the crossover under load will be correct. As the starting position of the tip relief is important to ensure correct deflection at crossover, we need a test which can measure the amount of tip relief without being masked by the meshing pair in-between holding the pure involute.

An answer to this masking problem is to increase the centre distance greatly so that instead of a contact ratio slightly over 2 we have a contact ratio slightly over 1. This depends on the tolerant properties of the involute and is only relevant for well designed and manufactured gears where most of the profile follows a pure involute. Because the centres have been moved apart, the length of "pure" involute has been roughly halved.

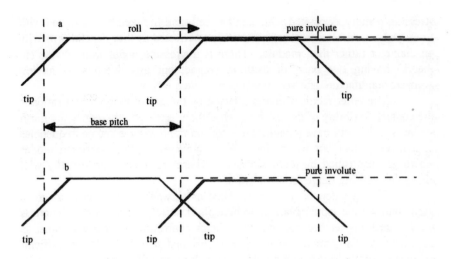

Fig 13.4 Sketch of T.E effects at (a) correct centre distance and (b) extended centre distance.

Fig. 13.4 shows diagrammatically the difference between two pairs of teeth meshing at correct and extended centres to give the same changeover points. The test requirement is then to decide what increase in the centre distance will give crossover points in exactly the same positions up the profiles of the gears as when handing over contact under loaded conditions.

The requirement is to find the exact positions up the profiles, not along the roll pressure line, where handover occurs for the original contact geometry and match these to the handover points at extended centres. This is an iterative calculation and it is simplest to use a computer routine to assist the process.

```
% program for finding centre distance change for contact ratio 2
% gears for TE for changeover. Work in terms of nominal module 1
phio = 18*2*pi/360 ; % design pressure angle 18 at contact ratio 2
n1 = 32 ; % number of pinion teeth
n2 = 131 ; % number of wheel teeth
br1 = n1*0.5*cos(phio); br2 = n2*0.5*cos(phio); % base radii
psi1 = (tan(phio) + 2*pi/n1) ; psi2 = (tan(phio) + 2*pi/n2) ;
% determine unwrap angles psi at changeover points assuming both
% are 1 base pitch away from pitch point
% these unwrap angles must be the same for extended test to be
% the same points on the flanks but will occur at roughly
% 0.5 base pitches away from the pitch point
```

```
% take first approximation to new pressure angle phi1 as due to
% centres moving 1 module apart so
phi1 = acos(cos(phio)*(br1+br2)/(br1+br2+cos(phio)));% new  angle
% then calculate distances from pitch point to changeover points
% divided by original base pitch
r1 = (br1*psi1 - br1*tan(phi1))/(pi*cos(phio));
% new pressure angle only original base radius real
r2 = (br2*psi2 - br2*tan(phi1))/(pi*cos(phio));
conratio = r1 + r2;
disp(' angle r1 r2   contact ratio ')
disp([phi1   r1 r2 conratio])                    % line 18
phi2 = input('enter new pressure angle   ') ; %  ****
r1 = (br1*psi1 - br1*tan(phi2))/(pi*cos(phio));
r2 = (br2*psi2 - br2*tan(phi2))/(pi*cos(phio));
conratio = r1 + r2;
disp(' angle  r1  r2     contact ratio ')
disp([ phi2  r1  r2 conratio ])
phi3 = input('enter new pressure angle   ') ; %  ****
r1 = (br1*psi1 - br1*tan(phi3))/(pi*cos(phio));
r2 = (br2*psi2 - br2*tan(phi3))/(pi*cos(phio));
conratio = r1 + r2;
disp(' angle  r1  r2  contact ratio ')
disp([ phi3  r1  r2  conratio])                  % line28
% calculate increase in centre distance from original
incr = (br1 +br2)*(1/cos(phi3) - 1/cos(phio)); % modules
disp('centre distance increase modules ')
disp(        incr )
```

The programme assumes that the original crossover points were placed symmetrically one base pitch away from the pitch point and calculates the involute unwrapping angles to these points. When the centre distance changes the only factors that remain the same are the two base radii and the two unwrap angles to the correct crossover points. The approach and recess distances after the centre change will normally not be equal. After the first guess at the new pressure angle only small changes are needed to adjust the angle (in radians) to give the contact ratio exactly 1.

If the original design was not symmetrical about the pitch point the original design values of the unwrap angles psi1 and psi2 to the crossover points should be used.

References

1.   Gregory, R.W., Harris, S.L. and Munro, R.G., 'Dynamic behaviour
        of spur gears.' Proc. Inst. Mech. Eng., Vol. 178, 1963-64, Part I, pp
        207-226.
2.   Leming, J. C., 'High contact ratio (2+) spur gears.' SAE Gear
        Design, Warrendale, 1990. Ch 6.
3.   Yildirim, N., Theoretical and experimental research in high contact
        ratio spur gearing. University of Huddersfield, 1994.
4.   Munro, R.G. and Yildirim, N., 'Some measurements of static and
        dynamic transmission errors of spur gears.' International Gearing
        Conf., Univ of Newcastle upon Tyne, September 1994.

14

Low Contact Ratio Gears

14.1 Advantages

Conventional industrial gears tend to use the standard 20° pressure angle and standard proportions and thus encounter undercutting problems when the number of pinion teeth falls below about 18. If gears are highly stressed they will normally be carburised and the standard AGMA2001 or ISO 6336 calculations will typically give a so-called "balanced" design at about 27 teeth. This means that there is an equal likelihood of failure by flank pitting or by root cracking. In practice as root failure would be disastrous, it is normal to have considerably less than 27 pinion teeth to make sure that root breakage is ruled out. This leads to most standard spur designs having between 18 and 25 pinion teeth and typically having a nominal contact ratio about 1.6.

Alternatively we can still get involute meshing with much lower tooth numbers if we are prepared to use non-standard teeth on the pinion. Tooth numbers of 13 or 11 are common on the first stages of small, high reduction gear boxes and the low tooth numbers allow larger reduction ratios. The designs use increased pressure angles typically of 25° and are "corrected" so the pitch circle is no longer roughly 55% of the way up the tooth but is only about one third of the way up the tooth when meshing with a large wheel.

For two equal gears meshing the practical limit is about 9 teeth and Fig. 14.1 shows two such gears in mesh. For pinion and large gear or the ultimate pinion and rack meshing the practical limit is down to 7 teeth. Again the pressure angle is 25° and the teeth are relatively narrow at the tips. The theorêtical contact ratio for these gears is about 1.05 to 1.1 but this nominal value does not allow for the relatively large contact area. Fig. 14.2 shows the geometry for a standard design which is used on oil jacking rigs where very large loads must be taken but pinion diameters must be minimised. These seven tooth gears with modules of the order of 100 mm (0.25 DP) are used with racks either 5" or 7" facewidth to lift the high loads of oil jacking platforms for use in waters up to several hundred feet deep. The loads on each tooth are then of the order of 500 tonnes and dozens of meshes work in parallel. Fig. 14.3 gives an expanded view of the contacts near the changeover point and it can be seen that there is very little overlap when there are two pairs of teeth in contact.

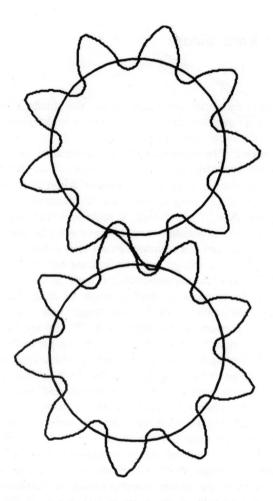

Fig 14.1 Shapes of two meshing gears with nine teeth and 25° pressure angle.

As can be seen in Fig. 14.3 with the contact ratio only slightly greater than 1, contact is occurring very near the pinion tooth tip and very near to the pinion base circle.

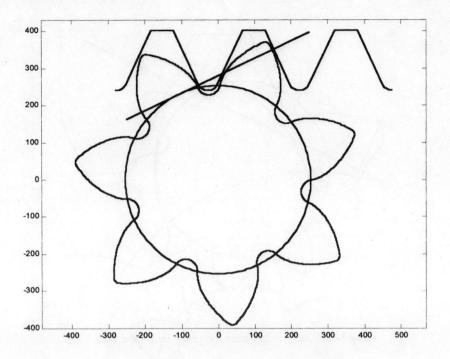

Fig 14.2 Seven tooth gear meshing with rack.

These highly loaded jacking gears work extremely slowly so noise is not a problem but stresses dominate the design. The major advantage in using only seven teeth is that the tooth size is dictated by the load carried. If the pinion were to have more teeth, not only would the pinion itself be larger and so much more expensive, but the driving torque necessary would be increased and so the cost of each drive gearbox would be greatly increased as cost is roughly proportional to output torque. Rather different considerations apply in the case of low power but high reduction ratio gearboxes. Here the main advantage of low tooth numbers lies in the reduced number of reduction stages and so less components such as bearings to be bought and mounted with the attendant costs. Less obvious advantages come from the more rapid reductions in shaft speeds so that there are fewer high frequency tooth meshes to rattle and give noise and there are fewer high speed shafts so lubrication and churning losses are lower. Lower tooth frequencies generally give lower noise.

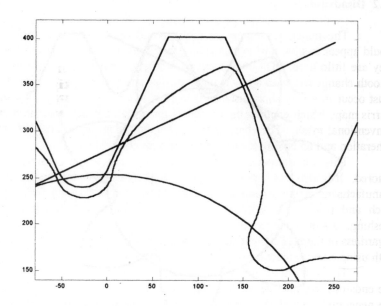

Fig 14.3 Detail of contacts for seven tooth and rack.

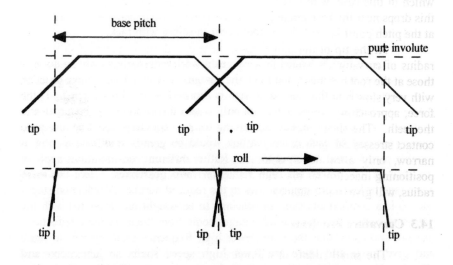

Fig 14.4 Contrast between tip relief shape for conventional design and corresponding fast change at tip for low contact ratio design.

14.2 Disadvantages

The major advantages in root strength associated with large teeth would appear to give low contact ratio gears a great advantage but in practice they are little used. The main reason for this is that it is difficult to get a smooth changeover with a low contact ratio as any theoretical tip relief design must occur in a very short distance if there is to be low T.E. Fig. 14.4 shows Harris maps which contrast the tip reliefs for a high contact ratio mesh and a conventional mesh. The changeover is very dependant on accuracy of profile generation and on having the centre distance exact.

This is not important for very low speed gears where dynamics can be ignored. It is also less important for very small gears since for small gears the manufacturing errors become much larger in relation to elastic deflections and pitch and profile errors become sufficiently large that they dominate the meshing. As the changeover errors are large they dominate the T.E. changes regardless of the nominal contact ratio so there is little noise penalty associated with using a low contact ratio.

The main disadvantages from strength aspects lie in the problems at the ends of the flanks where changeover occurs as exceptionally high stresses are generated. As can be seen in Fig. 14.3, at the bottom of the pinion tooth the contact is very near the base circle so the radius of curvature of the involute profile is very small. The standard Hertzian contact stress formulae for cylindrical contacts depend on the effective combined radius of curvature which in this case, with rack teeth, is equal to the local pinion curvature. As this drops near the base circle the contact stresses rise to about double the value at the pitch point.

At the tip of the pinion teeth there is a different problem in that the radius of curvature is relatively large so the Hertzian stresses are below half those at the root but the tips of the teeth are very narrow. There is high friction with very slow running gears so in one direction of rotation there can be a high force, approaching tangential in direction, attempting to shear off the tips of the teeth. The shear stresses across the narrow tip combined with the local contact stresses can give failure. Another problem can arise as the pinion tip is narrow, only allowing a small radius of curvature so manufacturing or positioning inaccuracies may run the contact onto the tip which, with its small radius, will give high contact stresses.

14.3 Curvature Problems

The small radius of curvature of the profile at the pinion root was mentioned as a problem in stressing. Our standard assumption is that the radius of curvature is equal to the length of the tangent from the base circle. In

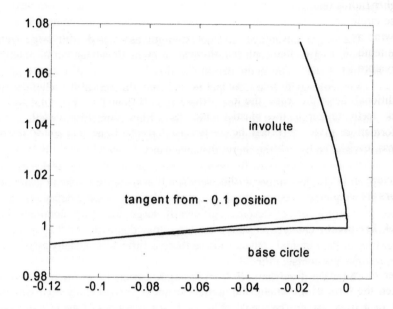

Fig 14.5 Expanded view of involute near base circle.

the limit, if the working profile reaches down to the base circle, the length of
the tangential unwrapping string becomes zero and then theoretically we have
zero radius of curvature and so very high contact stresses.

This does not agree with commonsense because if we look at the
shape of an involute as it starts out from the base circle, it does not look like a
small radius of curvature. It starts out by moving almost radially outwards as
can be seen in Fig. 14.3. with no hint of the sharp point we would expect with
zero radius of curvature. Double-checking the mathematics by alternative
methods still gives zero as the radius of curvature.

When mathematics and common sense do not agree it is usually
(invariably) the mathematics that is wrong. In this case the reason for the silly
answer is that near the base circle the centre of the radius of curvature (at the
tangent point to the base circle) is travelling as fast in the tangential direction
as the radius is reducing. The net effect is that the effective curvature is not as
sharp as expected. This presents problems when assessing contact stresses
since the effective radius of contact is very much higher than the theoretical
value.

Various attempts have been made to modify the involute shape near
the base circle to avoid the theoretical low radius problem but it is debatable
whether there is much point in such modification when there is in reality a

higher radius than expected. Fig. 14.5 shows the involute shape down near the base circle drawn out accurately and shows the tangent at the point 0.1 radian unwrap angle. With seven teeth this unwrap angle corresponds to only about one tenth of a base pitch. As can be seen, there is no detectable reduction in curvature for the first part of the involute.

For highly loaded gears such as jacking rig gears there is an additional factor that eases the local stresses. It is customary to design for the rack teeth to reach the plastic state each time they are loaded. The deformations involved spread the contact patch over a large area and so reduce stress levels greatly. The teeth surfaces deform permanently and the width of the rack teeth increases but the rack material is relatively soft and does not fracture and the required life of the gears is a restricted number of cycles so the gears are satisfactory.

14.4 Frequency gains

As mentioned previously, a standard "fix" for noise problems is to alter the number of teeth to alter the excitation frequency. This has usually taken the form of increasing the number of teeth to push the tooth frequency out of a troublesome resonance region. There is a stress penalty associated with finer teeth as root stresses rise and, in general, this approach will only help if the tooth frequencies are already high, say above 1 kHz. In general, reducing the size of the teeth does not reduce the T.E. at 1/tooth so it is equally likely that noise will rise.

An alternative that can be useful is when the 1/tooth is relatively low, say below 500 Hz. Reducing the number of teeth will drive the frequency down to the region where human hearing becomes much less sensitive and this is reflected in the standard A weighting used. At a given sound pressure level reducing the frequency from 200 Hz to 100 Hz corresponds to a nearly 10 dB improvement on the A weighting scale.

Another advantage of reducing frequency is that sound pressure levels depend on velocity of panel vibration so that if the vibration is at constant amplitude (as the T.E. remains constant amplitude) the frequency reduction reduces velocity correspondingly.

Speeds must be relatively low for frequency reduction to help. The standard motor speed of 1450 rpm will give about 400 Hz with a 19 tooth pinion so the number of teeth needs to be reduced to about 11 to pull tooth frequency down to the order of 200 Hz. If possible it is much quieter to use the traditional design of a 3 or 4 to 1 initial reduction by belt drive; then tooth frequencies are in a quiet region.

15

Condition Monitoring

15.1 The problem

Condition monitoring of gears (and of bearings) using vibration is an area where very large amounts of sophisticated electronics and computing mathematics have been employed at great expense but with rather limited effectiveness.

The objective is to give some form of warning of trouble before it happens, not after teeth have disappeared. This may be simply to allow industrial machinery to be maintained during the weekend before it breaks down and stops production in mid-week or, more critically, it may be to give the time necessary for a helicopter to land before the rotor jams. Alternative methods such as chemical analysis of the oil or debris monitoring are sensitive but tend to be too slow for immediate warning.

Originally, a couple of generations ago, standard accelerometers were fitted on bearing housings and a meter indicated rms or power over the whole frequency range. An overall rise in vibration power indicated trouble. The first development was to filter (analog) into octave or third octave bands and monitor the power in each band. The next stage (once cheap fast digital FFT routines were available), was to carry out a full frequency analysis, giving major lines at 1/rev, 1/tooth, etc., and watch each individual line. Any significant increase in amplitude of any line indicated trouble (in theory).

Some 30 years later a paper by Ray [1] summed up the state of the then current art. The vibration signal was frequency analysed but was also split into frequency bands, possibly six, covering the range, and each filtered band was subjected to a Kurtosis analysis. This involved taking the 4th order of the variation of the vibration signal from the mean (zero) and normalising it by dividing by the square of the mean power in the signal. The resulting non-dimensional statistical ratio would be less than 3 for a well-behaved random Gaussian distribution signal but would be greater than 3 for a signal which was "peaky." (In some work 3 is subtracted from the value.) The resulting criteria from frequency analysis line changes and filtered band Kurtosis figures were assessed to see if anything had changed "significantly" or if Kurtosis was too high and if so, red lights appeared to indicate that there was a fault. By the time a warning appeared it was often too late and a considerable number of false warnings destroyed operator confidence. Since then there has been considerable refinement of the electronics but, in terms of fundamentals, little progress.

It should perhaps be commented that gears are not usually the weak spot in gearboxes and that commonly it is bearings which fail, so any monitoring system must be good at detecting bearing problems. The requirements for bearing monitoring are surprisingly different from those of gear teeth but fortunately, monitoring bearings is, technically, a rather easier problem.

15.2 Not frequency analysis

The automatic reaction of a vibration engineer is to do a frequency analysis of a signal, but though this may be useful for noise (and may tell you how many teeth there are on the gear) it is of very limited use for damage monitoring. This is because FFT analysis gives the power in a spectrum line, spread over the test length which is usually 1 rev.

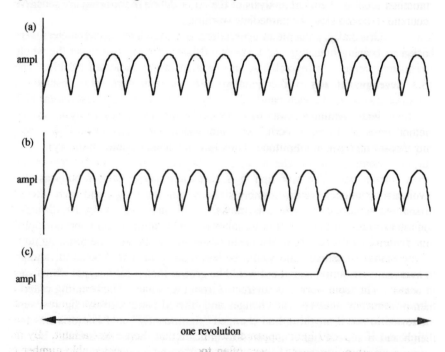

Fig 15.1 Time traces: (a) is for a single high tooth and (b) for a single low tooth; (c) is the difference of either from the regular 1/tooth pattern.

Fig. 15.1(a) shows an idealised signal, predominantly at 1/tooth for a revolution of a gear with an odd fault on one tooth and Fig. 12.1(b) shows a similar odd fault. Frequency analysis of such signals would show a negligible difference from the analysis of a gear with regular once per tooth and some background random noise.

In fact the difference between the results from either Fig 15.1(a) or Fig 15.1(b) and a regular waveform would be exactly the same as the frequency analysis of the subtracted signal shown in Fig. 15.1(c). A small pulse such as this, occuring for only a short time in the revolution would give very small components spread over a wide frequency range. These would be completely lost in the background noise and random variations present in any real system.

We are left with the problem that although we can see a fault very clearly in the original time trace, simple frequency analysis completely hides the fault so we will not see significant variations in line amplitudes unless all the teeth are damaged. This would be an extremely unusual or extremely powerful fault.

The same fundamental problem occurs with methods based on statistical analysis. Since a problem on 1 tooth of a 100 tooth gear may only occur for 1% of the time the power level associated with the problem is very low when spread over the whole revolution so it can easily disappear into the background noise.

15.3 Averaging or not

Time averaging of a vibration signal is a very useful and powerful method for reducing the volume of information and eliminating random noise and non-synchronous vibration. In general, it is useful for monitoring purposes but should be used with caution for some faults.

If a fault gives a perfectly consistent effect from revolution to revolution, then averaging is a great help. A hole in a gear tooth surface due to spalling or loss of part of a tooth will, in theory, give a signal which is consistent over many revolutions and which can be detected and analysed much more effectively if averaging at the frequency of that shaft is being used.

Wear or scuffing are by their nature inconsistent and not so amenable to averaging. A particular asperity that is being scuffed away may be removed in a few revolutions once the surface has been torn up and the scuffing may then move to another part of the tooth occuring at a slightly different time in the revolution. The effect of averaging over a large number of revolutions will then be to smooth out the variations over a long period and to hide the effects.

This leaves a problem in that, for monitoring cracking, major pitting or spalling, we might wish to average over a large number of revs, perhaps 256. In contrast, for scuffing, probably averaging over 8 or 16 revs would be more suitable so that we get some noise reduction effects but do not risk losing relatively transient effects.

A further possibility arises if we are interested in using vibration monitoring as a method of detecting dirt or debris passing through the mesh. Here the vibration pulse only occurs once or perhaps twice and we are interested in catching that part of the signal that is not regular. The most sensitive approach is to time-average the signal at both pinion and wheel frequency and subtract the averages from the original time trace (before averaging) to leave just the intermittent transient effects. An alternative is to high-pass the signal to remove eccentricities, then to average at once-per-tooth frequency and deduct to remove the main part of the "regular" signal, leaving mainly transients. With the main regular (low frequency) components removed, any short transients should be easier to detect. This will only work well if the once-per-tooth components are consistent.

15.4 Damage criteria

Starting from a vague feeling that damage ought to give some sort of variation on a vibration or noise signal does not give a direct indication of what an observed change of vibration means in terms of damage. It is worthwhile attempting to predict what character of signal the three standard types of damage might produce and how large that signal may be.

Pitting is the most common and widespread damage that occurs with gears and although 90% of pitting stabilises and is not threatening to gear life, it would be helpful to be able to detect it. On a medium-sized gear a pit may be 1 mm diameter. On a spur gear tooth with standard 20° pressure angle and 100 mm pitch radius at 1500 rpm the rolling velocity near the pitch line (where the pits usually occur) is 0.1 sin 20° * 50 π which is roughly 5 m/s. Assuming a working facewidth of 100 mm and a mean contact loading of 280 N/mm (20 μm elastic deflection) means that the expected change in force level will be at most 280 N if speeds were high enough that the gear masses did not have time to move. Alternatively, if the gears were rotating at very low speeds we would expect a displacement of 0.2 μm. This, of course, assumes that there is no averaging out of effects due to a thick oil film.

At full speed we may, at most, expect a differential force pulse [as in Fig. 15.1(c)] which was 280 N high and 0.2 milliseconds long. A half sinusoid pulse of this size would produce a displacement of a 14 kg mass of the order of less than 1/4 of a micron amplitude. This hypothetical size of displacement pulse must be considered in relation to normal T.E. excitation of the order of 5 μm. If there are several pitting craters near the pitch line the situation becomes more complicated since one pit crater may take over as another finishes, giving a relatively steady length of line of contact on a helical gear and, hence, a steady deflection.

A further complication arises with helical gears if we guess that there might be 20 pits associated with each tooth interval since our tooth frequency might be 600 Hz (1500 rpm and 24 teeth) and the pit frequency would then be 12 kHz. This high a frequency will be attenuated by the internal dynamics and will have difficulty in travelling out to the bearing housing accelerometers through either rolling bearings or plain bearings, even if the pulses are short enough not to overlap and give a steady deflection. Both hydrodynamic and rolling types of bearing tend to reflect fast pulses rather than transmitting them.

The overall conclusion is that it is going to be extremely difficult to see vibration effects at a bearing housing due to pitting. Part of the problem is that the excitation is small compared with normal T.E. and part is that high frequencies, well above internal system natural frequencies, will have very great difficulty in getting out to the bearing housing.

Tooth root cracking is potentially a very serious fault so it is worthwhile guessing what effect a cracked tooth would give. The main effect of a large crack along the root of a tooth would be to reduce the bending stiffness of the tooth and reduce the load taken by that part of the tooth. An extreme case would be if the stiffness was so low that the cracked part of the tooth took no load at all, as if that section of tooth had disappeared.

If we take a particular condition where 25% of the axial length of a helical tooth has "disappeared" then, with a contact ratio of 1.5, assuming perfectly even bedding along the total contact line length, the remaining contact line length will be 5/6 of the uncracked length.

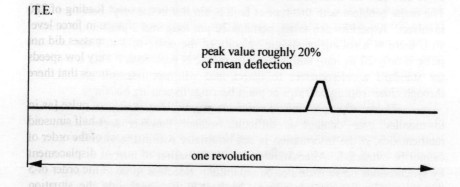

Fig 15.2 Change in T.E. due to part of tooth missing.

Ignoring system dynamics, this will give an increase of 20% in the mean deflection and would increase elastic deflection from say 20 μm to 24 μm.   The effect of this "missing" tooth section on static T.E. under full load will be as indicated in Fig. 15.2. This is a sketch of the change in T.E. that would be superposed on the normal T.E.

There would be a gradual run-in of the extra 4 μm with the rate depending on the exact design and a corresponding gradual runout. Since the changes are smooth there would not be very high harmonic components. This order of level of change, 4 μm, would be detectable in a very high precision gearbox such as a helicopter gearbox which was heavily loaded.

However, on a normal industrial gearbox, variations of this level can be encountered routinely from manufacturing errors such as pitch errors and, in position in equipment, there may be external transients as well as system dynamics to mask any effects from the broken tooth. High speed gearboxes present an additional problem because at 6000 rpm the tooth frequency is already 2.5 kHz and vibration pulses less than 0.1 milliseconds long are unlikely to be transmitted effectively through the bearings to the bearing housings.

Again the discouraging conclusion is that it may be difficult to detect much change in the vibration pattern even with a quarter of a tooth missing unless conditions are very favourable. This conclusion is borne out in practice since it has been known for significant chunks of teeth to be lost without any noticeable external effects. The damage was only detected when stripping down for routine maintenance.

The third main category of trouble is in the area of scuffing and wear, either due to breakdown of the oil film or due to debris and dirt in the oil. Metal-to-metal contact is involved and either asperities on the mating surfaces come into contact through the oil film or welding occurs between the surfaces. The major problem with this type of fault again lies in the very short time scale involved. Asperities are small, perhaps 20 μm long and 2 μm high, typically, so if there is a rolling or sliding velocity of the order of 1 m/s the pressure pulse is only 20 μs long and is typically only 2 N peak force. This is too short for standard accelerometers to detect and will not transmit satisfactorily through either rolling bearings or plain bearings to bearing housings.

These rather pessimistic estimates give an idea of why using vibration to monitor gear damage is difficult, because however sophisticated the mathematics, if the information is not originally within the vibration trace it cannot be extracted. Alternatively if the information of interest is dominated by synchronous noise it cannot be separated. Needless to say, any suggestion that the damage information may not be there in the signal early enough (to be extracted) is highly controversial with commercial developers of monitoring equipment.

The disturbances for pitting are essentially of the order of 1% of the mean load and are at too high a frequency to transmit out well. For root cracking, the disturbances are larger, typically of the order of 5% of the mean load but are comparable in size with commercial errors and in the same frequency range.

15.5 Line elimination

Since we are looking for small intermittent changes in pattern, a different technique is needed.

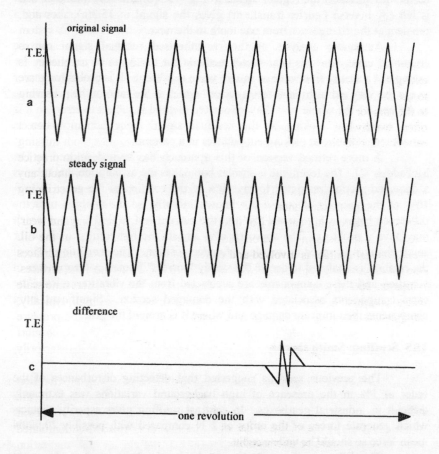

Fig 15.3 Line elimination and resynthesis to detect small changes.

In section 15.2 it was commented that frequency analysis would not easily show the small differences between Fig. 15.1(a) or 15.1(b) and a steady signal, despite the visible difference. A technique to show the difference is line elimination and resynthesis. This was mentioned in section 9.7, where the objective was to dispose of large lines but is of more general use to show occasional changes.

The example given in section 9.7 was of a small phase change on one displaced tooth but the same technique can show up other small changes. Fig. 15.3(a) shows a vibration trace for one (averaged) revolution of a gear.

FFT analysis followed by removal of all lines at 1/tooth and harmonics subtracts the regular signal in Fig. 15.3(b) and resynthesis of what is left (by inverse Fourier transform) gives the signal in 15.3(c), showing a problem at the changeover from one tooth to the next.

Automatic analysis of the resynthesised residual signal can be attempted using Kurtosis (statistical) methods but these can be unreliable, for example if a steady sine wave or square wave is present. It is probably simpler to use the very old fashioned "crest factor" which is the ratio of the peak value to the rms for the whole rev. Any automated method is subject to errors so it is often worthwhile looking at the residual signal since human vision is remarkably effective at picking out oddities in a pattern.

A more refined version of this approach has been developed by Dr. McFadden [2]. The technique is similar but makes the assumption that there is a damaged section restricted to say 10% of the rotation of the gear. Which 10% of the gear, is of course not known initially, so the analysis uses the difference between a previous test and the current test to find a sector which may have a problem. The approach also adjusts the test results to allow for small changes in speed or angular reference position. The remaining 90% of the rotation is analysed to derive the steady "correct" frequency components of vibration and these components are subtracted from the vibration to leave the extra components associated with the damaged section. Significant extra components then indicate damage and where it is around the gear.

15.5 Scuffing: Smith shocks

The previous sections suggested that detecting disturbances of the order of 1% in the presence of high background variations was extremely difficult in industrial gearboxes. In contrast scuffing gives asperity contacts which generate forces of the order of 2 N compared with possibly 20,000N mean force so should be undetectable.

The difference lies in the time scales involved since root cracks would give vibration at frequencies of the order of tooth frequency whereas scuffing gives pulses only about 20 μs long.

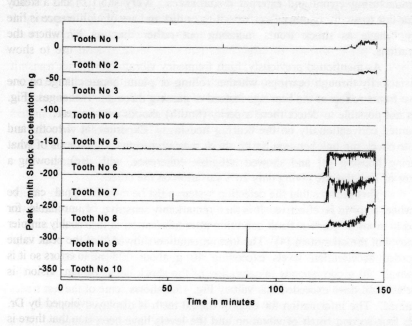

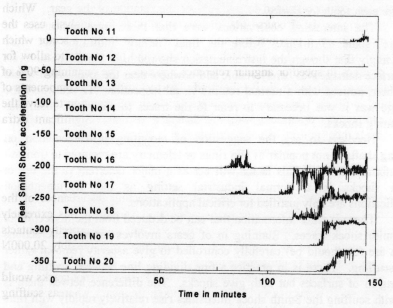

Fig 15.4 Test results from Smith shock investigations of scuffing failure.

Root crack vibrations are submerged in similar ones generated by manufacturing errors and external disturbances. Very short shocks do not occur due to anything other than asperities or dirt and are of high frequency so they behave as shock fronts radiating out rather than as lumped mass vibrations.

As mentioned previously, high frequency vibrations will not transmit satisfactorily through bearings, whether rolling or plain, because the pressure wave fronts reflect at the bearings instead of passing through. This means that it is not possible to detect these asperity (Smith) shocks using accelerometers mounted conventionally on the bearing housings. Experiments were carried out to check the link between Smith shock measurements on a rotor and on the bearing housing [3] and showed neligible coherence, with variations by a factor of 7 on the one giving only 20% variation on the other.

This dictates that the detection system must be mounted on the pinion or wheel rotor to be effective. It is then remarkably sensitive. The system was used to monitor gearbox flank condition after lubrication failure in the form of removal of the oil system [4]. The instrumentation showed "failure" with local recorded acceleration levels exceeding 40 g about 125 minutes after oil removal. 40 g corresponds to saturation of the shock detection system so the levels would have exceeded this value. Fig. 15.4 shows some of the test traces obtained. The information for each of the 20 teeth is displayed separately for each four second batch of vibration and the levels have been staggered down 40 g for each tooth for clarity.

The interesting observations were that there were indications of "failure" some 45 minutes before the final "failure" indication but more surprisingly that though the instrumentation showed high Smith shock levels the surface damage was barely visible and nowhere near the level which would have been noticed with a normal routine visual inspection. To see where the damage was it was necessary to refer to the traces to see which tooth was generating shocks.

Needless to say the suggestion of mounting instrumentation on rotating shafts is not popular as slip rings or telemetry are required to transmit the information out. This factor will act as a major deterrent to the use of Smith shocks in a normal industrial setting as the instrumentation complications are only justified for critical applications.

There is other information that can be deduced from the inspection of the Smith shock traces. Running in of gears involves asperity interactions which are (or should be) carefully controlled to give asperity removal rather than scuffing. There is in practice a fine dividing line between scuffing and running in of surfaces but both give shocks. The difference betwen them is that with scuffing the Smith shock intensities rise relatively rapidly with time whereas with running-in the shocks decay with time. They do not however decay to zero but to a level dictated by the residual surface roughness after the running in process. Monitoring the shock levels during running in allows a direct check on the progress so that the next stage of running in can proceed as soon as there is stability. Most running in procedures are uncontrolled and rather inefficient with much time wasted on regimes which are doing nothing of use.

Another slightly unwanted byproduct of monitoring Smith shocks is that, accidentally, they are the most sensitive and fast acting system ever encountered for detecting debris or dirt in the lubricating oil. Provided the debris is larger than about 3 μm every single particle passing through the mesh will generate a shock which is easily detected. Initial attempts to view scuffing on a 80 mm centres test rig were subjected to high "noise" levels from dirt in the oil. Elimination of this background noise for easy viewing of scuffing involved using clean oil which was filtered down to about 2 μm or better.

The major difference between debris shocks and scuffing shocks is that debris passage through the mesh only occurs once and so averaging over say 64 revs will virtually eliminate it whereas scuffing occurs in the same position each revolution for a small but finite number of revolutions.

The final conclusion is the slightly surprising one that although pitting and root cracking are almost impossible to detect in a normal accuracy industrial gearbox it is relatively easy to detect scuffing or to control running-in under laboratory conditions. Whether the very high sensitivity of Smith shocks to dirt in the oil will justify their use for monitoring debris in critical installations remains to be seen.

15.6 Bearing signals

Monitoring rolling bearing vibrations presents less practical problems than the corresponding gear vibrations. This is mainly because any vibration generated has a direct path to bearing housing accelerometers so there is no problem of lack of transmission of high frequencies.

If we compare signals from pitting with those from a damaged ball bearing track then in both cases we have a contact running over a pit which is typically about 1 mm across. There is a big difference in frequencies as pitting may involve a pulse which is only about 0.1 millisec long and of the order of 1% of the mean load whereas a track pit may generate a pulse some 5 millisec long with an amplitude of the order of 10% of the mean load.

In both cases there are characteristic frequencies or intervals between pulses involved which assist identification. Fig. 15.5 shows the typical trace obtained in one revolution but the next revolution, though it has the same time interval between pulses will have the pulses in a slightly different position as the cage speed is not synchronous with inner rotation speed.

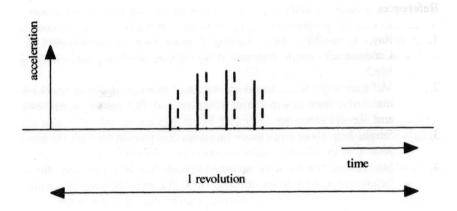

Fig 15.5 Expected trace from ball bearing with single inner track pit. Pulses shown as dashes are for the next revolution.

In practice the main problem with rolling bearings arises if the damage is not detected in the initial stages when there is only a single pit. Once damage has spread over a significant arc of the track the vibration signal generated is roughly continuous and the characteristic pulses disappear into a general background noise. In ball bearings any ball surface damage gives a rather intermittent signal as the ball tracks over different parts of its surface.

The standard techniques of frequency analysis and monitoring the amplitudes of the ball rotation and passing frequencies work well and will usually give clear warning of trouble. Roller bearings tend to present more problems as individual pits generate small pulses (as each pit carries a small fraction of the total load) and so generates a smaller pulse.

One unusual case occurs with fluid coupling drives which may be fitted between electric motors and gearboxes to cushion startup as, although these are running at normal speeds of 1450 or 1750 rpm, the internal bearings are only running at slip speeds of the order of 20 rpm. The relative speeds are so low that track or ball damage does not generate significant vibrations so it is almost impossible to monitor these bearings. In many such cases the use of a fluid coupling is redundant so it is preferable to remove the coupling and to rely on the protection systems for the motor to protect the gearbox as well as preventing motor overheating. The motor systems need to have the normal thermal (slow acting) cutout but also to have a current overload cutout which comes into action after the motor is up to speed. Soft start controllers can achieve the same result.

References

1. Ray, A.G., 'Monitoring Rolling Contact Bearings under Adverse Conditions.' Conference on Vibrations in Rotating Machinery, I. Mech. E., Sept. 1980, p 187.
2. McFadden, P. D., 'Detection of gear faults by decomposition of matched differences of vibration signals.' 2000 Mechanical Systems and Signal Processing. 14(5) pp 805-817
3. Smith, J.D., 'Transmission of Smith shocks through rolling bearings.' Journal of Sound and Vibration, Jan. 1995. 181 pp 1-6
4. Smith, J.D., 'Continuous monitoring of Smith shocks after lubrication failure.' Proc. Inst. Mech. Engrs., Vol 209C, 1995, pp 17-27.

16

Vibration Testing

16.1 Objectives

It may seem strange to think of deliberately vibrating a gearbox when an operating gear drive can be one of the most powerful vibration exciters that we have. The contrast between the excitation due to poor gears [which can easily give up to 10,000 N (1 Ton) p-p at tooth frequency] and the 45 N (10 lbf) from a typical small electromagnetic vibrator is rather extreme. There are several possible reasons for using an external forcing.

(i) Variable frequency. Many gear drives can only run at synchronous speed since they are driven by mains A.C. motors with low slip. We could get variability with modern three-phase inverter drives but this is quite a major change in the setup. The cost of inverters has dropped so much that this approach is now much more popular.

(ii) Cost. Running a large gearbox under load can waste a great deal of energy if the output is dissipated and setting up a back-to-back test rig is a major operation or may not be possible if there is not another gearbox available. As testing can take several days it is expensive to have to use high power test rigs.

(iii) Control and accuracy. The principle of general cussedness says that when we wish to have a meshing drive with a large regular excitation at 1/tooth and harmonics, there will not be a suitable "bad" gear pair available. Complications of modulation and variability under test will prolong testing.

(iv) Repeatability. Alignments and accuracies in a gearbox are temperature sensitive so it cannot be guaranteed that the gear excitation on a cold Monday morning is the same as on the previous Friday afternoon when everything was well warmed up.

(v) Ease of analysis. Having the input directly available for the transfer function analyser greatly simplifies testing.

The objectives of vibration testing are to find out as much as possible about the dynamic responses, both of the internal resonances and the external resonances. Testing external resonances is fairly straightforward, testing internal resonances is nearly impossible.

When the gearbox is running, the first place at which we can get a reliable vibration measurement is normally the bearing housing. When static (non-rotating) testing, if we attach a vibrator and generate an acceleration at a bearing housing, it does not matter to the gearcase, the supports, and the surrounding structure whether the vibration was generated internally by the gears or externally by an electromagnetic vibrator. Amplitudes will be much smaller with the (weak) vibrator but the structure and support system is assumed to be reasonably linear, so this should not matter. The setup can be as sketched in Fig. 16.1. In practice, the main excitation direction will usually be in the direction of the thrust line of the gears so we usually only test in this direction.

A conventional dynamic test running through the frequency range will show immediately whether or not there are casing or support frame resonances in the trouble area which is usually tooth frequency or harmonics.

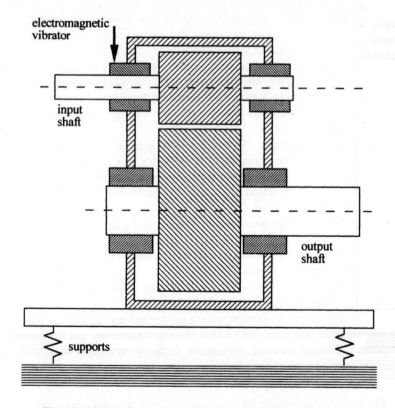

Fig 16.1 Setup for testing with electromagnetic vibrator.

If there is a resonance we need the mode shape and, as usual, running round the structure with a hand held accelerometer will give a quick check on whether there are individual panels vibrating excessively. As usual, panels where the centre sections are vibrating more than the support points, [as in Fig. 10.1(c)] must be tackled, whereas panels with reduced amplitude [Fig. 10.1(a)] are working well.

One question with such a response test is what the "output" should be. Using a noisemeter at a standard position is useful and is the obvious final arbiter but the sound measured may be affected by direct sound radiated from the vibrator itself.

There is no obvious place to put an accelerometer, so if the suspicion is that the important transmission is through to the structure, placing the accelerometer on a mounting foot might be the best position. Otherwise it requires an iterative process to find the "worst" (i.e., the highest) amplitude position and then use that point as the standard reference point.

A possible alternative is to both excite and measure at a bearing housing so that the local vibration response is obtained. The results of such a response test are sometimes presented "upside down" as the force per unit acceleration and the result is called an "effective mass" of the system.

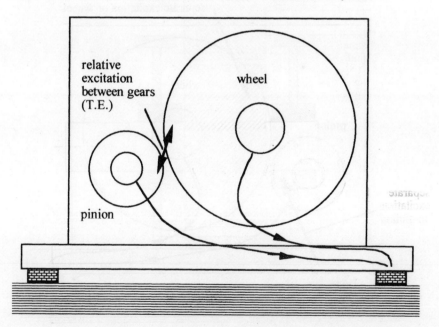

relative excitation between gears (T.E.)

wheel

pinion

Fig 16.2 Multiple paths for vibration via the four bearing housings.

Results from such a test can be rather confusing as high sound production can occur both at a resonance where amplitudes are high or at an anti-resonance where power input is not high, but a panel is resonating to oppose motion in the manner of a tuned absorber. Either low mass (resonance) or high mass (anti-resonance) can then indicate a problem area. As the results are confusing it is better to scan round the structure and look for high amplitudes.

Reality is slightly more complicated than a simple test, as indicated in Fig. 16.2. The original excitation from the gear teeth will transmit to the casing at both pinion bearings and both wheel bearings so, in theory, we should vibration test at the four bearing housings, setting up amplitudes and relative phases to correspond to running conditions.

This would be far too complicated so testing at one pinion bearing is normal. We assume that wheel vibrations are likely to be sufficiently smaller to be ignored as stiffnesses and inertias of the wheel are large. The alternative is to carry out a full linear sensitivity analysis using the responses from each of the four bearing housings, but the effort is not justified.

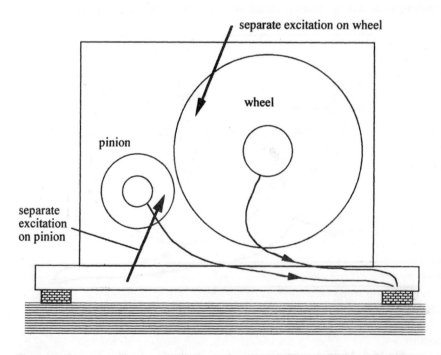

Fig 16.3 Separate vibrator excitations on gears so that the vibrator mass does not affect the results.

In general, we only need to vibrate once at a given bearing housing in the direction of the line of thrust as this is the dominant excitation. It is not too important to get the excitation absolutely correct if the main objective is to locate natural frequencies and determine the important (noisy) mode shapes.

Internal resonances are almost impossible to test directly. The ideal would be to have a vibrator small enough to fit between the meshing gear teeth and giving a relative displacement to excite the system, much as T.E. does. Since we only have available the backlash space of about 100 μm (4 mil) this is unrealistic, but if we use two opposed exciters on the gear bodies we are not incorporating the all-important contact meshing stiffness between the teeth.

The alternative is to excite the gear bodies separately as indicated in Fig. 16.3 and then add the results and the estimates of the tooth effects as discussed in section 16.6.

16.2 Hydraulic vibrators

As electromagnetic vibrators are large and very heavy yet low on force, one possibility is to use hydaulic vibrators as commonly used for testing machine tools under realistic conditions. A double-sided ram is fed at high pressure through a fast servo valve. The exciter is very small and light (less than half a kilogram) so it can be fitted into cramped spaces, with the accompanying 80 gallon tank and drive unit at a convenient distance. As pressures are high at 200 bar (3,000 psi) the vibrating ram need only be about 12 mm effective dia. to give a force of 4,500 N (1000 lbf) p-p. Such a vibrator is small enough to fit into a machine tool but not into a gear mesh.

One problem is that hydraulics do not like high frequencies. The servo valves used have a flow of the order of 5 gal/min (0.3 l/s) and a natural frequency of about 220 Hz. At this frequency they will drive a 12 mm ram at a maximum speed of about 3 m/s corresponding to an amplitude of ±2 mm. This assumes a good design with high flow areas but negligible dead volumes and with no conventional seals in the main ram section as they are too elastic. To a reasonable approximation the vibration is limited by maximum flow.

220 Hz is towards the lower end of the audible range and by the time the frequency has risen to a more characteristic 700 Hz the flows have decreased by a factor of 10 so the maximum possible amplitude of vibration is down by about 30 and so below 100 μm. A stiffness for a gearcase might be 10 N μm^{-1} so the force generated would be 1000 N but dropping off fast with frequency. In practice attempting to work over 1 kHz is not worthwhile.

The combination of the expense and noise of a hydraulic system together with the problems associated with high frequencies tend to rule out the use of hydraulic vibrators. There is the additional problem that phase lags in the servo valves are high, of the order of 1° per Hz so it is difficult to control the system and open loop operation may be needed. For this it is advisable to have a combined system where the average force is controlled (at very low frequency) by the slow servo loop while the vibrating force component is set by hand.

16.3 Hammer measurements

Electromagnetic vibrators are clumsy, delicate and expensive and hydraulics is cumbersome so an alternative is desirable. The method which has become much more popular with improvements in computer processing is the use of impact testing.

Mathematically a pure Dirac impulse has infinite force amplitude for zero time but with a finite area (usually 1) under the impulse and this gives harmonic components which are at all possible frequencies and are of equal amplitude which is zero. This is a rather abstruse concept but fortunately we are only interested in the transfer function between the input (force) and the output vibration. As long as the pulse is short enough to have frequency components covering the range of response we can have any shape of pulse. We then take a frequency analysis of the input and of the output and the ratio (complete with phase) is the frequency transfer function required.

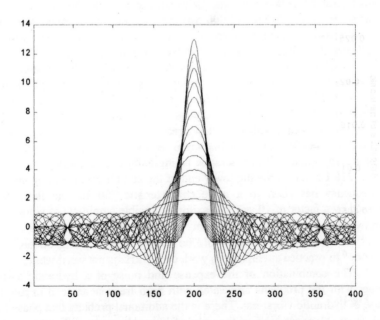

Fig 16.4 Build up of harmonics to give a pulse.

Having too short an impulse is not helpful as too much of the input energy goes into high frequency components which are above the response range of interest. Usually the length of the impulse can be adjusted by altering the contact tip material and the geometry.

The idea that an impulse will give all harmonics at roughly equal amplitudes is rather strange to some students as they are accustomed to the idea that all frequencies and equal amplitudes in a vibration give random "white noise". The only difference between the two is that the phases of the components in white noise are completely random whereas for an impulse the phases all coincide (at zero) at the pulse. Fig. 16.4 shows the first few (13) harmonics adding together at one point to build up an impulse.

The rough relationship to give an idea of how long a pulse is needed for testing is indicated by the result shown in Fig. 16.5 which gives the frequency content for a half sine pulse of unit height and length 4 milliseconds. It can be seen that if the pulse length is τ then by frequency $1/\tau$ the amplitude has decreased to 40% of the value at low frequency. Most of the energy in the impact has occurred by frequency $1/\tau$.

Using a soft plastic tip to increase the impact time is sometimes not feasible as it will deform under the load which is normally well over 1000 N.

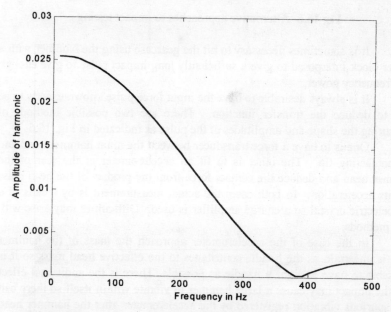

Fig 16.5 Frequency distribution of input for 4 millisec half sine pulse.

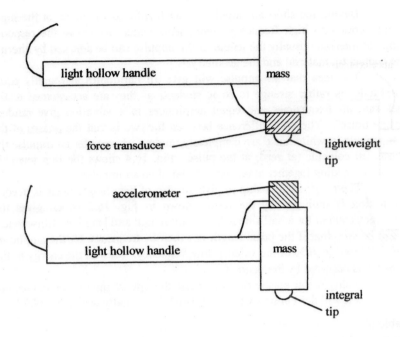

Fig 16.6 Alternative approaches to impact testing.

It is sometimes necessary to hit the gearcase using the hammer with a rubber block interposed to give a sufficiently long impact pulse to give enough low frequency power.

It is always desirable to have the input force pulse known exactly to be able to deduce the transfer function. There are two possible methods of measuring the shape and amplitudes of the pulse as indicated in Fig. 16.6.

One is to have a force transducer between the main hammer head and the contacting tip. The other is to fit an accelerometer at the rear of the hammer head and deduce the contact force from the product of the head mass and its acceleration. In both cases the actual measurement is by means of a piezoelectric crystal so a charge amplifier is used. Difficulties may arise with both methods.

In the case of the accelerometer approach the mass of the hammer head is uncertain as the handle contributes to the effective head mass so it is advisable to have as light a handle as possible. There is the additional effect that the impact may cause a large hammer to vibrate within itself so there can be a spurious vibration registered by the accelerometer after the hammer head has lost contact with the target.

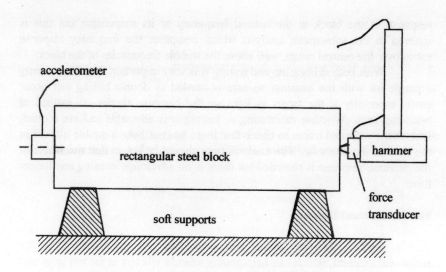

Fig 16.7 Use of steel block to calibrate hammer.

Using a force transducer would appear to be a much more direct and reliable method but this does not seem to happen in practice. There are some theoretical effects reducing the effective sensitivity due to accelerating forces on the hammer tip but the disparities between theoretical and observed force sensitivities are larger than expected. It is not unusual for the output from the force transducer to be less than 70% of the value quoted by the manufacturer unless the hammer head is very heavy.

As one of the reasons for using an impact approach is to have highly portable equipment this is a limitation. With uncertainty existing about force transducer effective sensitivity it is advisable to carry out a calibration with the hammer to be used. Calibration, whether of a force transducer type of hammer or an accelerometer hammer involves using a block of known mass on a soft suspension.

The acceleration of the block is measured by an accelerometer of known sensitivity and preferably by the accelerometer which is going to be used for the tests on the gearbox. Accelerometer calibrations supplied by the manufacturer are in practice very reliable and will give accurate readings of vibrations.

The test mass should be of the same order of mass as the expected gearcase response effective mass and so might be 100 kg for a large drive or 1 kg for a small drive. The simplest setup is to mount the steel block on a very soft foam rubber support or on large soft rubber bungs as indicated in Fig. 16.7 and to impact horizontally but care is needed to hit exactly in line with the centre of gravity of the block and the accelerometer axis. There will be a response of the block at the natural frequency of its suspension but this is ignored in the subsequent analysis which compares the frequency response ratios over the central range, well above the wobble frequencies of the block.

With both calibration and testing it is very important that there is only a single hit with the hammer so care is needed as double hitting can occur easily especially if the target is light so the hammer carries on instead of bouncing back. Whether calibrating or testing it is advisable to have a quick look at the recorded traces to check that there has not been a double hit within the time of the analysis. The analysis time should be set so that the whole of the gearcase response is recorded but there is no advantage in using any longer time.

16.4 Reciprocal theorem

One of the very useful approaches we can use comes from the reciprocal theorem, which can help greatly when a vibrator is far too large and bulky or it is not possible to get a clear hammer swing at a bearing.

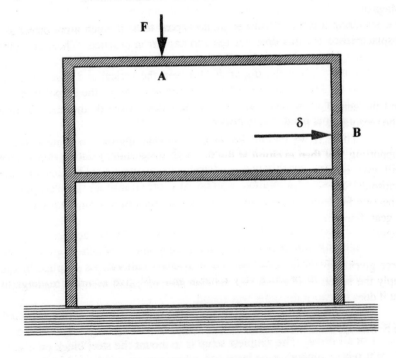

Fig 16.8 Reciprocal theorem in a structure.

The rough rule is that the force from an electromagnetic vibrator is roughly the same as the dead weight of the vibrator and though hydraulic vibrators are relatively small and compact for their power they do not work satisfactorily at 1 kHz. As electromagnetic vibrators are large, heavy and delicate (as well as being expensive) they cannot be mounted inside a gearbox and sometimes are too large to fit near bearing housings but the resulting vibration near an accessible foot may be wanted.

Fig. 16.8 shows a simple static application of the reciprocal theorem in a structure. If a load F applied at A gives a deflection δ at B, then a load F applied at B (in the same direction as δ) would give the same deflection δ at A (in the direction of F). This is extremely useful in structural and vibration analysis and is roughly intuitive for a simply supported beam.

This result is a simplified (static) case of the more general property that if we apply a vibrating force at A, the dynamic response at B will be the same as it would if the force were applied at B and the response measured at A.

The static case was deduced by Maxwell and Betti in 1872 and Rayleigh extended it to the dynamic case in 1874 in the Philosophical Magazine. The responses in both amplitude and phase will be the same. In the standard response terminology the receptance (acceleration, velocity or displacement per unit force)

$$\beta_{AB} = \beta_{BA}$$

and the complex response can be described either in r, θ or in $x + jy$ form. The receptance is the inverse of the (complex) stiffness.

Thus, if we cannot excite at a bearing housing and measure at a supporting foot then exciting at the foot and measuring at the bearing housing will give exactly the same result. This is equally true whether we excite sinusoidally or with an instrumented impulse. The same applies if we excite at a bearing housing and measure at a gear flank because we cannot get access to a gear flank for a vibrator (but can for an accelerometer or laser velocity meter).

It is important that the reciprocal theorem is used correctly so that force gives displacement (or velocity or acceleration) not the reverse. We can apply the superposed forces and the displacements will then add (vectorially) but it does not work the other way round.

16.5 Sweep, impulse, noise or chirp

When vibration testing to get a transfer function or response of a system there are four basic choices:

(a) the traditional very slow sweep with a vibrator,

(b) noise (band limited) with a vibrator (it is not necessary to have pure white noise but all frequencies in the range should be present),

(c) fast sweep (chirp) with a vibrator, or

(d) using an impulse which requires an instrumented hammer.

Since they give the same result apart from some small experimental differences, it is not too important which method is used. The slight differences obtained with impulse are due to the absence of a back reaction from the vibrator base but only very low frequency modes are affected. At high frequencies the added mass associated with an electromagnetic vibrator can alter natural frequency. Choice is more a matter of familiarity, convenience and the availability of equipment, than for any fundamental reasons.

Methods (b), (c) and (d) must have some form of twin channel transfer analyser, usually in computer form for economic reasons. In contrast (a) can be done completely by hand, using an oscilloscope. Slow sweeps do not fit well into the standard twin channel transfer function approach which data-logs for a limited time. Method (d) has the great advantage of not involving the practical problem of mounting a clumsy vibrator. The peak force is much higher, but the power in a given frequency band is no higher than with a slow sweep vibrator so repeated impulses are needed for averaging to give improved accuracy. As a rough rule the power at any given frequency when using an impulse is a factor of 30 down on the peak force so to match a vibrator which delivers 60 N force we should impact with a peak force of the order of 2000 N. It requires manual skill to get a reliable result and to get consistency, while the stiffness of the hammer tip has to be adjusted to get the best frequency range.

White noise, containing all frequencies simultaneously, appears to give much faster testing than sweep or chirp methods but, because peak vibrator force is limited, there is very little power in any single frequency band. In a normal, slightly noisy, environment this means that the testing must be run for a long time or there must be multiple tests for averaging to achieve a reliable result so there is negligible gain in overall testing speed. Chirp methods are a compromise between slow sweep and white noise methods but are now little used because they require rather too long a signal for conventional fft analysis.

Once the important response frequencies have been located and mode shapes are required, the approaches differ because a vibrator cannot easily be moved, whereas an impulse hammer is easily moved. A vibrator is kept fixed (at the bearing housing) and an accelerometer moves round the gearcase or structure to determine the deflected mode shape, usually with the vibrator at a single (trouble) frequency for maximum signal-to-noise ratio. This method, once the vibrator has been set up, is fast and the information is easily understood. In contrast, with impulse testing, the reciprocal theorem is invoked.

The accelerometer is left fixed at the bearing housing and the hammer is moved round the structure. Setting up is much easier but the information extraction is slower and it is not so easy to make mode shape deductions from the results. The great advantage of the single frequency vibrator method is that the information is immediately visible to the operator so it is much easier and faster to understand mode shapes in very fine detail when there are complicated local distortions in a structure.

16.6 Combining results

The simplest case of combining results occurs at a foot of a gearbox where we can measure vibration when running on a test rig. We can measure the dynamic response at the foot to an external vibrator and we can measure the structural (hull) dynamic response.

For simplicity, assume that the test rig has extremely soft antivibration mounts so that the gearbox feet are not restrained on the test rig and the foot vibration is then the "free" vibration level.

Fig. 16.9 shows an idealised setup with foot F and hull H and, to complicate matters, an intermediate elastic isolator E. All vibrations and forces are in the same (vertically downward) direction. We can then call the various responses (or receptances) β with suffixes denoting where we excite and where we measure. All the receptances are complex to allow for phase effects. When the gear drive is running "free" on soft test supports, the vibration is of amplitude V and we wish to predict the hull vibration δ and the force P transmitted through the mount when the gearbox is installed.

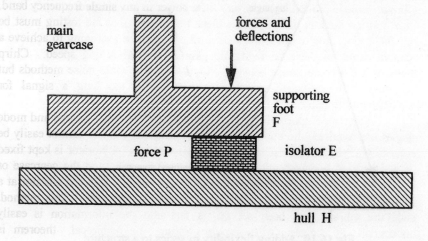

Fig 16.9 Idealised conditions at gearbox foot.

The first move is to decide what the internal forcing force is and we use superposition to find out how much force would need to be supplied to stop the foot vibrating. This is then the internal driving force.

With local response β_{ff} (measured with a vibrator) and amplitude V, this force is V/β_{ff} so this is the equivalent internal forcing force P if driving against a foot held rigidly. At the gearbox foot we now have a net force $V/\beta_{ff}-P$ acting and so the response will be $V - P \beta_{ff}$. At the hull, the force P will give a response $P \beta_{hh}$. The difference between the foot deflection and the hull deflection will be the isolator deflection $P \beta_{ee}$ where β_{ee} is the inverse of the (complex) stiffness of the isolator. Then

$$V - P \beta_{ff} - P \beta_{hh} = P \beta_{ee}$$

so we can find P as $V/(\beta_{ff} + \beta_{hh} + \beta_{ee})$ and, hence predict the hull vibration. Since the various responses may be nearly out of phase they may roughly cancel each other and give resonances at some frequencies. This, as described, assumes that the effects at one foot are independent of excitations from hull forces at the other feet. Although this is not true, it gives a good first approximation and avoids the major complications of cross effects from one foot to another. As far as the (vector) additions of forces and displacements are concerned, it does not matter whether we are measuring displacements, velocities or accelerations.

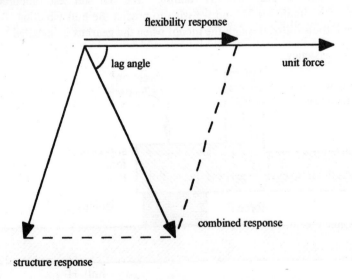

Fig 16.10 Adding flexibility in series to a structure.

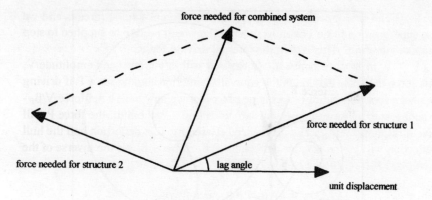

Fig 16.11 Connecting two structures together.

It is sometimes much easier to work with vector diagrams to see the effects of connecting two systems. In general it is preferable to use diagrams when only one frequency is being investigated and to use complex responses when the whole frequency range is being considered.

Fig. 16.10 shows the effect of connecting an extra flexibility such as a tooth stiffness or an elastic isolator in series with a structure and Fig. 16.11 shows determination of the result when we connect two structures together so that they are working in parallel.

In the one case force is common and we add the displacements and when we fix two structures together the displacement is common and we add the forces. Care must be taken with the directions of forces and displacements to ensure consistency.

It is more complicated when we have excitation at the gear teeth as in Fig. 16.12, with a force between the teeth of F. The wheel displacement at position 1 is due mainly to the force F acting on the wheel at 1 but is also due to the force F acting on the pinion at position 2. So

$$\delta_1 = F\beta_{11} + F\beta_{12} \qquad \text{and} \qquad \delta_2 = F\beta_{12} + F\beta_{22.}$$

As the transmission error is the sum of δ_1 and δ_2

$$\text{T.E.} = F\,(\beta_{11} + 2\,\beta_{12} + \beta_{22}) \quad \text{giving } F.$$

Once F is known, the displacement δ_3 at the support foot can be found as

$$\delta_3 = F\,(\beta_{13} + \beta_{23}), \quad \text{etc.}$$

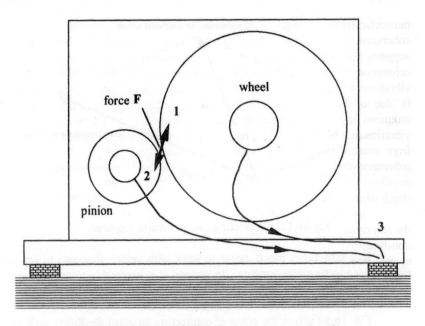

Fig 16.12 Sketch of vibration responses and paths.

In practice, because measuring inside the gearbox is very difficult, it is probably better to rely on T.E. excitation when running or to estimate the internal resonances. The reciprocal theorem is of limited help since, although it may help for cross receptances β_{13} and β_{23}, it does not assist access for β_{12} or the local responses β_{11} and β_{22} which need access inside the box in zero space. In some cases the wheel is so massive and its support is so stiff that wheel response may be ignored, simplifying the algebra considerably.

16.7 Coherence

Whichever method (b), (c) or (d) is used for measuring a transfer function with a transfer function analyser, it is worthwhile checking coherence if there is any possibility of background noise, whether mechanical or electrical. The idea of coherence is that if we take a single transfer function measurement we can deduce a transfer function. However, we do not know how much of the output is really due to the input and how much is due to random external (or internal) disturbances.

Repeating the test many times and getting exactly the same result in both amplitude and phase would suggest that there is little random effect. Any variation would suggest randomness. Coherence analysis routines carry out this check and compare how much of the measured output power (at a particular frequency) can be attributed to the consistent transfer function. A coherence of 1 suggests that output is firmly connected to input but < 0.5 suggests that random noise is dominating the measurements. Any results with coherence < 0.8 should be viewed with suspicion. Even if there are two vibrations whose coherence is 1 it is not necessarily true to say that the output is "due to" the input since both vibrations may have been generated by another unknown input. In particular a panel vibration may not be caused by the vibration at a bearing housing because both may have been caused by vibration from another bearing or even from a separate slave drive. To carry out a coherence check it is necessary to take multiple tests, typically eight. It is not possible to get a meaningful result from a single test because it is necessary to check whether the result is consistent over time.

Extra care should be taken when impact testing because even though the responses may be consistent from test to test there is a greater likelihood of non-linearity. This in turn will lead to false deductions since a high response at one frequency may in fact be due to excitation at, say, one-third of the frequency encountering a non-linearity.

17

Couplings

17.1 Advantages

Couplings in a system are rarely fitted initially with a thought to their effect on noise. The most common requirement is that the coupling must accommodate misalignments which may be due to manufacturing or assembly errors but are often due to the effects of differential thermal growth, which can be surprisingly large. A temperature difference of 40°C can easily occur between a turbine casing and a gearbox and if the centre height is 1 m the corresponding differential expansion is 0.4 mm (16 mil). Axial growth can also present problems since a motor gearbox combination in a medium-sized (400 kW) installation with a distance of 4 m between foundation attachment points and a temperature rise of 50°C can expand axially by 2 mm.

As far as noise is concerned, the use of a coupling is usually advantageous since those couplings which use rubber blocks as the torque transmitting units have flexibility for both torsional and lateral vibrations. The steel diaphragm type of coupling, usually used in pairs with a short torque shaft in between, is torsionally stiff but laterally flexible.

Toothed gear couplings are short and light and have lateral flexibility and, in theory, axial flexiblity but like diaphragm couplings have high torsional stiffness.

In most installations the transmission of gear noise along the input or output shafts is not important as there is likely to be a large inertia for load or driver and so vibrations will be absorbed by the inertia. Typically this occurs on a car where any torsional vibrations from the gearbox encounter either the high inertia of the engine or of the wheels. In the case of the wheels there is also the filtering effect of the propshaft flexibility to attenuate vibration.

The exceptions to this occur when there is a large propellor or turbine which can act as a very effective radiator of noise. On naval ships the propellor has a large surface and the vibration frequencies are high so that any vibration will radiate powerfully and betray the ships position. Under these circumstances it is critical that some form of very flexible coupling is used for isolation of both lateral and axial vibration.

A similar requirement occurred recently with the installation of wind turbines for "renewable energy" purposes. Early designs did not consider that noise would be a problem as the installations were meant to be away from dwelling houses. They made the mistake of connecting the propellor directly to a gearbox which was chosen for low cost rather than low vibration. The result was that the relatively large vibrations at 1/tooth and harmonics were transmitted straight through to the propellor. This acted as a remarkably efficient loudspeaker with a very large surface area and produced a gear whine which could be heard miles away. The eventual solution was not only to use high quality gears and reduce propellor dynamic flexibility but to isolate the propellor from the gearbox with a soft rubber torsional coupling. In addition, of course, the gearcase had to be effectively isolated from the supporting tower which could also act as a noise radiator. At the other end of the drive there was no problem as the high inertia of the generator absorbed all vibration very effectively.

17.2 Problems

The problems associated with rubber couplings are usually at low frequencies where either the torsional flexibility of the coupling gives a torsional resonance at a frequency too near the running speed or the mass of the coupling brings whirl speeds down into the operating range. This effect on whirl speed can of course also occur with diapraghm couplings.

It is difficult to carry out accurate predictions because the properties of rubber couplings are not well documented. This is partly due to production variations which give a surprising spread of rubber hardness which can vary some ± 20% so that it is possible to find a "soft" unit which is stiffer than a "hard" unit of the same design. In addition the characteristics of the filled natural rubber which is usually used vary at low amplitudes both in the stiffness and the damping factor as well as varying with frequency. Typically dynamic stiffness may be 40% higher than the figures quoted by the manufacturer (as they are given for low frequency response).

Reliable information can be obtained by using a back-to-back rig to give an exact replica of operating conditions as indicated in Fig. 17.1. High drive torque is applied statically, then the intermediate ring is oscillated torsionally at the correct (low) level using two opposed exciters mounted tangentially and is measured using two tangential accelerometers. It is necessary to mass correct for the moment of inertia of the intermediate ring.

Torsional couplings, like conventional vibration isolators, may also have been designed and installed with the main objective being to isolate 1/rev and 2/rev vibration and so may be ineffective for the much higher frequencies of gear noise.

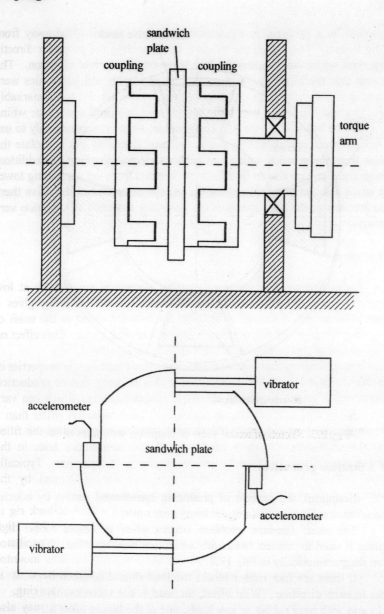

Fig 17.1 Back-to-back test rig for torsional stiffness under working torque. The high static torque is applied by the torque arm, which is then locked.

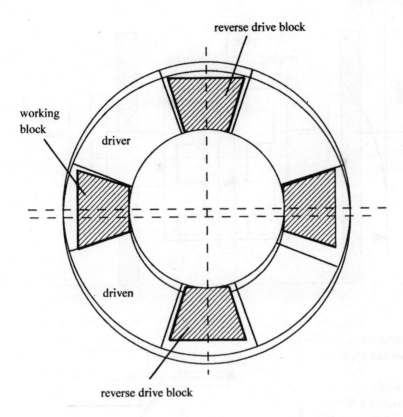

Fig 17.2 Sketch of axial view of coupling with axes offset.

17.3 Vibration generation

Couplings are capable of producing unexpected results by injecting torsional excitation or modulating existing gear noise.

The most common problem occurs when a simple rubber block coupling is used to connect two shafts which are slightly offset. The effect is shown diagrammatically in Fig. 17.2.

If there are four rubber blocks the load should be taken by two of the blocks in each direction. With offset, the load is not taken evenly by the two blocks and with hard rubber or low loads, one of the blocks takes all the torque and there is clearance on the other block for half of the rev then the other block drives for the other half of the rev. The resulting error is as shown in Fig. 17.3 and with an amplitude peak-to-peak equal to the offset acting at block radius. Alternatively manufacturing tolerances can give once per rev errors.

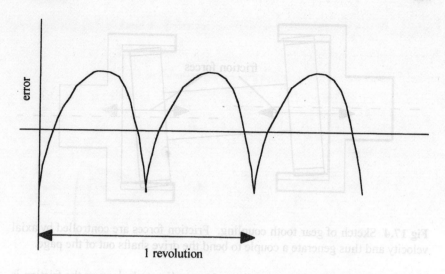

Fig 17.3 Torsional transmission error with offset.

Assembly of a drive motor onto a worm and wheel gearbox with conventional tolerances can easily involve an offset and runout of 100 μm and this can appear as either a 1/rev or 2/rev effect according to the type of error.

Any attempt to measure T.E. under these circumstances will give (at 1/tooth or 2/tooth) an apparent gear error of the order of the offset and so possibly a factor of 10 larger than the true gear errors. This effect can occur to a limited extent with higher numbers of blocks but will be small if all the blocks are under sufficient load to be in contact all the time so that the system remains linear.

Diaphragm couplings are preferably radially symmetric and so will not inject torsional vibrations into a drive but the trailing link type of coupling needs to have more than two links to be self centering and so to be satisfactory when used at each end of a torque shaft.

Gear tooth couplings can produce some very unexpected results. They are radially symmetric and so we would expect a smooth drive with no injection of extra frequencies. They will in practice run vibration free when they are perfectly aligned and also when they are badly misaligned. Vibration problems can arise when they are only slightly misaligned.

In a gear tooth coupling there are friction forces as sketched in Fig. 17.4 and there will also be some bending elasticity in the drive shafts. There are two extreme cases.

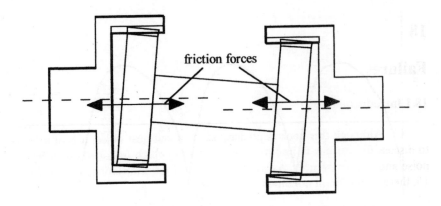

friction forces

Fig 17.4 Sketch of gear tooth coupling. Friction forces are controlled by axial velocity and thus generate a couple to bend the drive shafts out of the page.

Near perfect alignment allows the coupling to lock up as the friction is sufficient to bend the shafts so that there is no relative axial motion at the meshing gear teeth and there is no vibration excitation apart from a 1/rev bending on the shafts. Significant misalignment gives continuous sliding at the coupling gear teeth and thus no significant vibration injection. The problem arises with small misalignments which will initially bend the drive shafts because there is not sufficient force to overcome the axial friction at the teeth but after perhaps one-third of a revolution the friction will be overcome and there will be an axial slip at the teeth. This effect will inject a disturbance into the drive at 3/rev, altering the shaft bending at this frequency and so disturbing any neighbouring gear mesh at this frequency. This can lead to the modulation of the gear noise frequency so that noise occurs at tooth frequency plus or minus 3/rev. Deliberate alteration of the misalignment may change the slip frequency higher or lower or it may disappear completely.

As far as testing T.E. is concerned it is much safer to test the gear drive separately without any couplings in place to get the basic gear information then, if the couplings are suspect, to test the complete assembly. Unfortunately this must be done under load as otherwise the friction forces will not be correct.

18

Failures

18.1 Introduction

Although this book is predominantly about gear noise, it is of interest to discuss the various failure mechanisms to see which might be connected to noise and which are not. As may be deduced from the comments in Chapter 15, there is in general not much connection.

18.2 Pitting

Pitting arises from traditional Hertzian contact stresses giving failure as a result of a fatigue process. The standard theory [1] gives the results that for line contact, i.e., cylinder to cylinder with load P'/unit length, the maximum contact pressure p_0, and the semi contact width b, will be

$$p_o = \left(\frac{P'E^*}{\pi R} \right)^{1/2} \qquad\qquad b = 2 \left(\frac{P'R}{\pi E^*} \right)^{1/2}$$

Effective curvature $\quad \dfrac{1}{R} = \dfrac{1}{R_1} + \dfrac{1}{R_2} \quad$ where R_1 and R_2 are the radii of curvature.

Contact modulus $\quad \dfrac{1}{E^*} = \dfrac{1-v_1^2}{E_1} + \dfrac{1-v_2^2}{E_2} \quad$ where E_1, E_2 and v_1, v_2 are Young's

moduli and Poisson's ratio, and suffixes 1, 2 refer to the two bodies in contact.

The maximum shear stress is then $\quad \tau_{max} \cong 0.300\, p_0 \quad$ at $x = 0$, $z = 0.79b$

This leads to a maximum shear stress occuring typically about 0.5 mm below the surface and giving fatigue cracks which, for traditional pitting, travel initially horizontally then curve upward toward the surface. When they reach the surface a hemisphere of steel breaks out leaving the classical pit which is typically 1 mm diameter and 0.5 mm deep.

269

tip

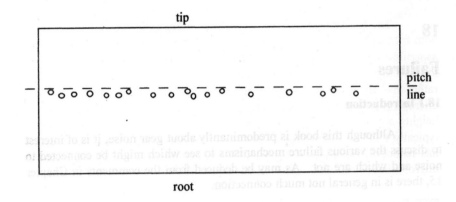

pitch
line

root

Fig 18.1 View of tooth flank with pitting.

The simple static theory suggests that pitting will be at its worst where stresses are highest because the effective radius of curvature is smallest, which is when contact is toward the root of the pinion but this is not what happens. The pitting occurs initially very near but not exactly on the pitch line where sliding velocities are low and the typical pattern is sketched in Fig. 18.1. In cases where a gear pair is significantly misaligned the pitting will concentrate on the highly loaded areas. The result is an area of metal removal which is sometimes called spalling [2]. Whether the term spalling should be used for this localised heavy pitting is debatable as it was formerly used for the rather different failure when the "skin" of an inadequately carburised gear peels off, giving an effect labelled as case/core separation in the AGMA 1010. The flake pitting [2], which is sometimes encountered, is similar and may also be caused by faulty carburising.

Pitting depends on fatigue and so is a relatively slow process which in most cases stabilises. Occasionally the loadings are too high for the material and the pitting progresses and covers the whole gear surface but even this serious deterioration is unlikely to produce "gear" noise because as mentioned in Chapter 15 the frequencies are very high and tend to be reflected or to be absorbed before reaching panels which could radiate noise.

18.3 Micropitting

Micropitting (sometimes called gray staining) has become more important recently, possibly as a result of greater use of case-hardened gears and changes in manufacturing techniques. It has similarities to conventional pitting but occurs on a much smaller distance scale and occurs at slightly lower loads than pitting. Unlike conventional pitting, it tends to spread and progress and may start anywhere on the flank.

The initiation is due to asperity contacts generating local high stresses with friction forces assisting the process. It differs from pitting in that because asperities are small, with sizes of the order of μm, the stress fields are very localised so the pits generated are comparable in size with the surface finish rather than the mm-sized stress fields of pitting. 20 μm is a typical depth of a micropit [2]. A prime requirement for micropitting to occur is that the asperity heights are of the order of, or greater than, the oil film thickness which is typically 1 μm or slightly less. The use of synthetic oils at high temperatures has tended to reduce oil film thicknesses and so increases the likelihood of micropitting.

As far as noise is concerned, the comments that apply to pitting are even more relevant. The scale of the micropits is so small that at normal running speeds the frequency of the pits is above the normal audible range so that even if the vibrations were transmitted they could not be heard. In practice they do not transmit out through the bearings.

There is currently considerable interest in micropitting but tests carried out as long ago as 1987 [3] indicated that using a mirror finish so that the surface roughness (about 0.1 μm) was less than the lubrication film thickness (0.4 μm) gave increased resistance to micropitting. This would then raise possible operating conditions to the normal pitting limit as dictated by Hertzian contact stresses. It is unfortunate that the standard grinding processes tend to leave a rather rough surface finish which encourages micropitting.

18.4 Cracking

Traditionally cracking occurrs at the tooth root as sketched in Fig. 18.2. The crack starts at a surface stress raiser somewhere in the tooth root, well away from the working flank and once started it spreads rapidly so that the complete section of tooth falls out. On a helical gear it is not usual for a complete tooth to fail but perhaps one-third of the width of the tooth may crack off.

This form of failure is very rare since it is liable to be rapid and disastrous. Because it is so serious, normally design carefully avoids it and the flank pitting should occur first. Tooth root cracking is usually an indication of faulty design or faulty heat treatment.

The surprising feature is that tooth breakage can occur and may not be noticed until a routine stripdown uncovers it. Noise generation is usually not noticeable and even monitoring equipment may miss it. The major hazard is if the broken tooth attempts to go through the mesh and jams the drive.

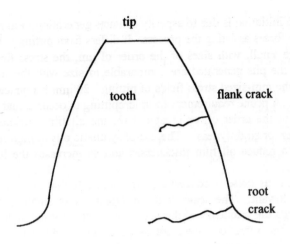

Fig 18.2 Crack positions.

Cracking can also give trouble starting in the middle of the working flank, often near the pitch line. The initiation in this case appears to be due to the cracks which (macro) pitting and micropitting generate and which may branch downward into the main body of the tooth instead of branching upwards to give a pit. Friction at the contact appears to play a significant part and this, in turn, is very dependant on the lubrication conditions. As with conventional root cracking there is likely to be little or no noise generation.

18.5 Scuffing

Scuffing involves breakdown of the oil film so that metal-to-metal contact can give welding and subsequent tearing and flow of the surfaces. It may be associated with too thin an oil or excessive loading or large sliding velocities giving too much heat input to the oil.

The curiosity in relation to scuffing is that the process is very similar to running in of gears. Both are associated with asperity contacts which result in metal removal and the main difference is one of scale. Running in removes the (small) asperities and the surfaces become smoother whereas with scuffing the scale is larger and welding occurs so the surface dragging gives rougher surfaces. The borderline between the two processes is not clearly defined and can only be followed experimentally by monitoring with Smith shocks [4]. A downward trend of the shock level indicates successful running in whereas an upward trend shows scuffing and the conditions should be altered immediately.

When the scuffing is due to lack of lubrication it is possible not only to halt the damage by restoring lubrication but to improve the surfaces to restore their full load carrying capacity. On some slow but very heavily loaded gears the use of a good grease can heal surfaces that have previously been damaged by cold scuffing.

As far as noise is concerned the initial stages of scuffing are very irregular and local to a single point in a gear so that there is not a regular pattern linked to 1/tooth or other expected frequencies in the audible range. The 1/rev impulses that are generated are short and so should give a noise similar to a burr or isolated damage on a tooth. Once there is a significant scuff the vibration can be detected by conventional accelerometer monitoring but the deterioration may be very fast by that stage.

18.6 Bearings

The normal pattern of design for gearbox bearings has been that only very high power gearboxes needed to use hydrodynamic bearings with their cost and complications of high oil flow rates needing hundreds of horsepower to achieve the cooling rates required. Hydrodynamic bearings might be needed simply because speeds were too high or because specific loadings were too high for rolling bearings.

Medium-sized gearboxes are now encountering loading limitations increasingly due to improvements in gear loadings and to the basic scaling laws for gears and rolling bearings. As a very rough rule the load on a gear may be increased proportional to size squared whereas the load on a bearing may increase less rapidly. If we take figures for the "heavy duty bearing", a spherical roller bearing, then within an O.D of 190 mm we get a C rating of 535 kN and an infinite life rating of 67 kN. Doubling the O.D. to 380 mm allows a C rating of 1730 kN and an infinite life of 193 kN. This seems to follow a rough rule that doubling the size increases capacity by a factor of three whereas on a gear we would expect an increase by a factor of four.

The corresponding gear size may be estimated very roughly by using the 100 N/mm/m rule. 20 teeth of 20 mm module with a "square" pinion gives a load of 100 x 20 x 400 which is 800 kN. For any long life installation such as a chemical works or sewage plant it is advisable to use the infinite life value so we find that a bearing of slightly less than the size of the pinion will only take one quarter of the gear load. The situation is slightly eased if the pinion support is symmetrical but the two bearings can only take half the possible gear tooth load. Taking the full load symmetrically with two bearings requires an O.D. of 460 mm and an asymmetric design with 600 kN load on one side would need 520 mm O.D. There would be enough radial space for this with a large wheel but not with a low reduction ratio.

The effect of this is to force the designers to use lower safety margins and hence increase the possibility of rolling bearing failures. It is unlikely that incipient failure of a rolling bearing will give audible noise problems but it is fortunate that this type of damage can usually be picked up effectively by bearing housing monitoring. Use of high loads makes bearings much more susceptible to dirt or debris as the oil films are thinner and the extra stresses due to the particles are imposed on stresses which are already high.

Other problems that may give bearing failure stem from there being insufficient load on a bearing. The manufacturers give empirical rules for estimating the minimum load at high speeds but these are for steady speeds only. Heavy torsional vibration of the sort associated with light loads and inaccurate gears (rattle) can demand high torsional accelerations of the rolling elements and if loads are light then skidding of the rollers can occur and damage the bearing rapidly. One heavy duty drive was unwisely tested under no-load and failed in a couple of hours but would have operated for many years under full load.

Multistage gearboxes with high reduction ratios present severe design problems since there may be, say, 100 to 1 variation in speed so the ideal oil viscosity for the high speed gears and bearings is totally unsuitable for the low speed gears and bearings. It is advisable to bias the choice toward the low speed bearings and increase the viscosity even though this will increase the lubrication losses and thus increase heat generation.

18.7 Debris detection

Traditionally debris detection is one of the oldest techniques for giving indication of trouble. Magnetic plugs were of limited use since they were usually only inspected when an oil change was scheduled. Modern particle counting techniques are very effective to give an accurate quantitative assessment of the state of the oil and must be used if bearings are heavily loaded and so are very vulnerable to dirt or debris in the oil.

The results of debris analyses are in the form of the number of particles counted in 100 ml of oil and the figures are surprisingly high. Several versions can be used but, for gears and rolling bearings, the two figures which are usually quoted (and are of most interest) are for the number of particles above 5 μm and the number of particles above 15 μm respectively. The numbers are not given directly but are classified on an approximately binary scale.

A brand new clean oil might be naively expected to be particle free but in practice may have a test count of 200000 / 7000 and so would be classified as 18/13. There are sometimes three figures quoted but then the first figure is for the particle count over 2 μm and is not of interest for gearboxes.

It is of interest to compare the test sizes of 5 and 15 μm with the expected oil film thicknesses, which range from less than 1 μm in a rolling bearing to 3 to 4 μm in a medium-sized but lightly loaded gear.

The particle counts are classified into groups according to ISO 4406 and the figures are

Particles per 100 ml			Group
500000	to	1000000	20
250000	to	500000	19
130000	to	250000	18
64000	to	130000	17
32000	to	64000	16
16000	to	32000	15
8000	to	16000	14
4000	to	8000	13
2000	to	4000	12
1000	to	2000	11
500	to	1000	10
250	to	500	9
130	to	250	8
64	to	130	7
32	to	64	6

Bearing manufacturers typically suggest that 18/14 is a "normal" cleanliness for oil and so the above new oil would be "normal".

Unfortunately, to be able to use rolling bearings to their full capacity "normal" cleanliness is nowhere near good enough and the requirement is to achieve much better. FAG suggest that 14/11 is needed but better cleanliness is needed if the contaminants are abrasive (such as sand). Work done at SKF [5] suggests that for rolling bearings, debris of the expected high hardness will give raceway damage when the particle size exceeds 5 μm. Below this size it appears that the elastic deformations of the surfaces can accommodate the particles without reducing the life. SKF suggest that for absolute maximum life the contaminants should be comparable in size to the oil film thickness but this means there should be very little debris above 1 μm. INA also state that for "extreme cleanliness" particle sizes should be less than the film thickness. Normal industrial practice is to have what the bearing manufacturers would classify as typical contamination with a heavy life penalty.

There is a further problem for the gearbox user in that there is no direct connection between the specification for the filter and the corresponding particle count in the gearbox. Filters are specified in terms of their reduction ratio for particles of a given size so that a $\beta_6 \geq 75$ as suggested by FAG would reduce particle count for those above 6 μm by a factor of 75 for a single pass through the filter. Whether a recirculation of the filtered oil will reduce the count by another factor of 75 is not discussed. The resulting contamination in the oil depends on how fast the oil is being circulated through the filter and, more importantly, how fast fresh contamination is entering the system (possibly from the gears?).

In practice the only reliable solution is to monitor the oil particle content. After a fresh batch of oil (at 18/13) is added to the system the particle count should drop to 14/11 or preferably better and should stabilise. Any subsequent increase requires a change of filter and probably a check on the source of the debris.

Gear contact oil films are thicker than rolling bearing films so should be less susceptible to dirt but as gears have some sliding rather than pure rolling motion there can be tendency to give scratching on the tooth flanks. This means that it is sometimes easier to detect debris in a gearbox by looking at scratching on the gear flanks than by attempting to see the inaccessible roller tracks where any debris damage will be at a point instead of producing a scratch.

When there are thick oilfilms as with plain bearings or the rolling bearings are lightly loaded it is not the bearing which is the critical member and a build up of debris will show up as abrasive wear on the tooth flanks. This is especially so with spiral bevel gears which have high sliding velocities and so are very vulnerable to dirt.

Looking at the various failure mechanisms and their likely effect on oil debris is not encouraging. Tooth root cracking produces one large lump which with luck will drop to the bottom of the gearcase and not move so there will be no indication of trouble from debris analysis whether chemical or particle counting. Pitting (macro) again produces a few relatively large hemispheres which will not show up in a particle count and will usually stay at the bottom of the oil tank or sump. Scuffing should produce some fine debris and so should be detectable but only micropitting would produce large quantities of relatively fine debris particles.

The conclusion is that surface wear (due to debris) or micropitting will put up particle counts but that the other failure mechanisms will have little effect. Any connection between debris and noise is unlikely as normal debris is small so gives pulses which are at too high a frequency to hear and which occur intermittently.

As mentioned previously in Chapter 15, the most sensitive debris detection system yet encountered for small particles is using Smith shocks to detect the particles passing through the mesh but this is too sensitive for use in normal commercial gearboxes and is unlikely to be used due to the experimental complications involved.

18.8 Couplings

Couplings appear to be an unimportant part of the system but can not only occasionally themselves fail but can produce failures in other parts of the drive.

Design for the steel diaphragm type of coupling is straightforward as the coupling stiffnesses, axially and in bending, should be given in the sales literature and so it is easy to predict what loadings will be applied to the shafts on either side. Axial loadings need to allow not only for assembly errors but also for thermal differential expansions.

The rubber block type of coupling is much used in small drives as it can accommodate some offset of axes as well as angular misalignment so only one coupling is needed instead of two with the diaphragm type. The corresponding disadvantage is that although the blocks deform to take the offset there is a significant sideways force which may fatigue shafts in bending. A rather unusual problem can arise due to thermal effects if the axial growth is sufficient to take up the clearance in the coupling. The metal (or plastic) castings can then meet and impose severe axial forces to produce failure of motor or gearbox input bearings.

Gear tooth couplings are compact and light and can take high torques so they are popular in high power drives. They are, however, able to impose considerable bending torques on their supporting shafts as mentioned in Chapter 17.

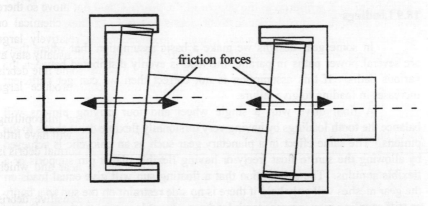

friction forces

Fig 18.3 Sketch of gear tooth coupling. Friction forces are controlled by axial velocity and thus generate a couple to bend the drive shafts out of the page.

In Fig. 18.3 assume a drive torque T, gear radii R and gear spacing 2R, then the tooth forces total T/R and if we make the pessimistic assumption that due to the tilt most of the forces are concentrated as shown by the double ended arrows then the bending moment at each end of the coupling is R x μ T/R, where μ is the coefficient of friction. At the centre of the coupling by symmetry there is no bending moment so there must be simply a shear force of μ T/R. This shear force acts to bend the drive shafts and if the overhang of the centre of the coupling from the centre of the supporting bearing is 3R then the resulting bending moment on the drive shafts is 3 μ T. Taking a value for μ of 0.2 gives 0.6 T bending moment and unfortunately this is an alternating moment which will attempt to fatigue the shaft especially if there are any local stress raisers.

An alternative effect is that this bending moment on the shaft may cause trouble by misaligning an input pinion or sun wheel of an epicyclic and so affecting the gear stressing by increasing the load distribution factors C_m and K_m. If a gear is tilted by this effect and the load is not evenly distributed along the facewidth the T.E. may be increased and so give more noise.

The other problem that can occur with gear tooth couplings is when alignment is good. The coupling then locks up in the same way that a spline locks up when torque is applied and can give high axial forces which may reduce bearing life. A gear tooth coupling with a gear diameter of about 100 mm will have a rated torque of about 3 kN m so the tangential forces at the gear teeth will be of the order of 60 kN and if the coefficient of friction after lockup is 0.16 there will be a possible axial load of 10 kN or 1 ton. This could easily destroy a gearbox input bearing.

18.9 Loadings

In some gear designs we make a basic assumption that where there are several power paths in parallel the load is evenly distributed between the various paths. If this assumption is not correct then we can get sufficient increases in loading to give failure.

A final drive with a single wheel and four driving pinions will balance the tooth loadings by having very torsionally flexible drive shafts to the pinions. The same effect in a planetary gear such as an epicyclic is achieved by allowing the sun to float freely or having flexible planet pin supports or a flexible annulus. The assumption that a floating sun will give equal loads on the gear meshes will only hold if there is no side restraint on the sun so a faulty or stiff coupling may increase tooth loadings. Input by a gear tooth coupling or stiff coupling as in the above section can unbalance loads.

An extreme case of unequal loadings can occur in installations such as oil jacking rigs where there can be 36 or 54 electric motors all working in parallel through reduction gearboxes to raise or lower 20,000 Tons by rotating pinions which mesh with vertical racks.

The assumption is made that loads are equal when designing the gears but it is relatively easy to imagine conditions where due to structure effects there is unbalance so that there may be 50% increase in the load on a single drive. It is advisable to design on the basis that this may occur and to take care in selecting the drive motors so that they cannot give too high a torque. Occasionally wiring errors occur so that one poor motor is attempting to drive downwards while the neighbouring seven are driving upwards. This tends to play havoc with the stressing but is difficult to design against. It is unusual for unbalanced loadings to have an audible noise effect so noise is of negligible help in detecting unbalance.

In general care is needed with unusual or new drives to ensure that loads are as expected. Early wind turbine problems arose due to frequent overloads of up to 70 or 80% due to gusts of wind. The hydraulic controls on the blade feathering could not respond fast enough to prevent these overloads so the drives failed. The other relatively common problem is with step-up drives where there is a high speed rotor with a large inertia. The step-up gearbox is subjected to full starting motor torque of perhaps 250% of design torque for a significant time each startup and so will fail unless designed for double torque. Alternatively a soft start motor control must be used although this carries the penalty of longer runup times.

18.10 Overheating

Cooling of gearboxes is rarely a problem in small sizes as surface area relative to power is high and in very large sizes there is usually an external cooling system to control the oil temperature.

Overheating may occur when natural convection is relied upon but the heat generation is greater than expected. The normal single stage reduction gearbox of about 5 to 1 ratio with 1450 rpm input will have the wheel running at about 300 rpm so that oil churning is restrained as only part of the wheel dips into the coolant. Natural convection can dispose of about 1 kW per m^2 of surface and this is usually adequate. Inserting an extra stage into the gearbox will not cause heat generation problems if the extra bearings and gears are running slowly but if a high speed shaft is added the churning losses will increase greatly and may give oil breakdown. It is then necessary to drop the oil level to prevent churning and add spray cooling directed at the gear teeth. Worst of all is to have a shaft running at high speed with rolling bearings and gear meshes completely immersed in oil.

At high speeds the use of external spray cooling will reduce the heating from the gears but there is a danger of the rolling bearings overheating unless they are well drained so that they are not running full of oil as this causes high heat production. A designer may be so concerned to get cooling oil into a bearing that he has the oil going in faster than it can get out. In critical cases it may be necessary to use oil mist cooling to get sufficient cooling without too much oil.

References

1.   Johnson, K.L., Contact Mechanics, C.U.P., 1985.
2.   ANSI/AGMA National standard 1010-E95. Appearance of gear teeth - terminology of wear and failure.
3.   Tanaka, S. Ishibashi, A.,and Ezoe, S. Appreciable increases in surface durability of gear pairs with mirror-like finish. Gear Technology, March/April 1987, pp 36-48.
4.   Smith, J.D., 'Monitoring the running-in of gears using Smith shocks.', Proc. Inst. Mech. Eng., 1993, vol 207 C, pp 315-323.
5.   Ioannides, E., Beghini, E., Jacobsen, B.,Bergling, G. and Goodall Wuttkowski, J. Cleanliness and its importance to bearing performance. Journal of Society of Tribologists and Lubrication Engineers, 1992, pp 657-663.

19

Strength versus Noise

19.1 The connection between strength and noise

It is often assumed, sometimes unconsciously, that a noisy gearbox is one that is likely to break. This comes from the observation that a gearbox that is disintegrating (usually because of bearings failing) becomes noisy (or noisier) and so noise is associated with failure.

Usually there is little connection between noise and strength and if a system keeps the gear teeth in contact it is rare for vibration to affect the gear life. The time when noise and strength are directly connected is when the teeth are allowed to come out of contact and then produce high forces in the following impact. High noise and high stresses are then both associated with the repetitive impacts as discussed in section 11.3.

The extreme cases where noise and strength give rise to dramatically different designs are:

(a) Ultra low noise teeth with a nominal contact ratio above 2 where the minimum number of tall slender teeth is above 25 and the pressure angle is lowered.

(b) Ultra high strength gears for lifting self-jacking oil drilling rigs where 7 tooth pinions mesh with racks at a pressure angle of 25 degrees.

This lack of connection between noise and strength presents difficulties when it comes to testing the gears on production. If we are targeting minimum noise then the only worthwhile test is a T.E. test but this is of no use for assessing strength. Conversely, if the requirement is for maximum strength, especially for low speed gears, then it is essential to carry out a bedding check to make sure that the major part of the face of the gear tooth is working but a bedding check is not a valid predictor of noise.

Production is left with the problem that they need to know whether noise or strength is the more important and if both are important, then both tests must be carried out. This is unfortunate because bedding is a relatively slow and expensive test. Skilled labour is required and the test is time consuming so costs rise. T.E. testing is very much less expensive but is rather unknown as yet in general industry so it is viewed with great suspicion and is avoided wherever possible.

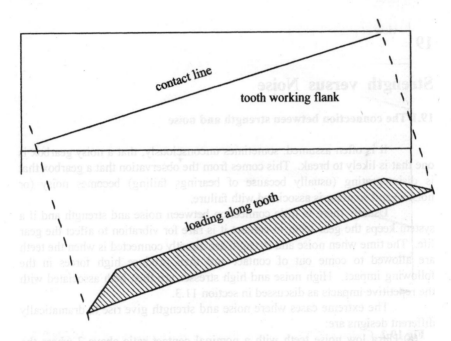

Fig 19.1 Desired loading pattern along contact line for maximum strength.

Much depends on the application since within a given gearbox (such as the previously mentioned car gearbox), there may be strength dominant on the two lower gears, requiring bedding checks, and noise dominant on the three higher gears, requiring T.E. checks.

19.2 Design for low noise helicals

From a "philosophical" aspect it is relatively easy to design for maximum strength. If we look at a helical gear flank as in Fig. 19.1 we need to get the maximum length of line of contact, compatible with reducing the load to zero at the ends of a line of contact. Within the line of contact, the objective is to get the loading per unit length constant over the length of the line.

This objective results ideally in a trapezoidal shape to the loading distribution along the length of the line of contact. There is little choice in the resulting "ideal" design, apart from how fast we reduce the loading at the ends. It is preferable to use end relief instead of tip relief to maximise the area of full loading or to use "corner" relief if the extra manufacturing cost is justified. However, to achieve a good loading across the facewidth the helix alignments must be extremely good, to within, say, 20% of the mean tooth deflection.

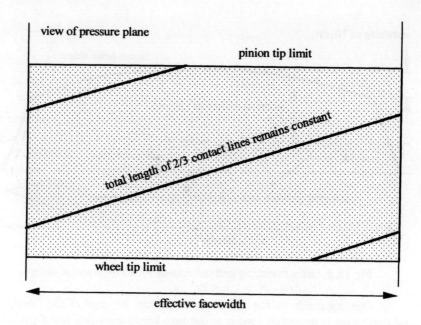

view of pressure plane

pinion tip limit

total length of 2/3 contact lines remains constant

wheel tip limit

effective facewidth

Fig 19.2 Contact lines when facewidth is an exact number of axial pitches.

In designing for low noise there are more options available and much depends on whether or not there is a good margin of strength in the design.

If we could rely on perfect helix alignment, life would be fairly simple since, apart from tip relief and end relief needed to prevent corner loading, we could use virtually any profile at low load.

At high load, if the axial length of the gear is an exact number of axial pitches then the contact lines on the pressure plane would always have the same total length. This is shown in Fig. 19.2 and would give constant mesh stiffness, hence constant elastic deflection and a smooth drive. Such a design is, of course, also a high strength design if there is negligible relief at the tips or ends.

Unfortunately, the reality is that helix alignment is very rarely better than 10 μm (0.4 mil) and the error is more likely to be much greater, of the same order as the theoretical elastic tooth deflection. This end loading not only puts the load concentration factor across the facewidth (C_m or $K_{h\alpha}$ x $K_{h\beta}$) up above 2 or even 3, but prevents the helix effects from averaging out the profile effects. We are left with the necessity of assuming that the helix alignment will be poor and thus need to design accordingly.

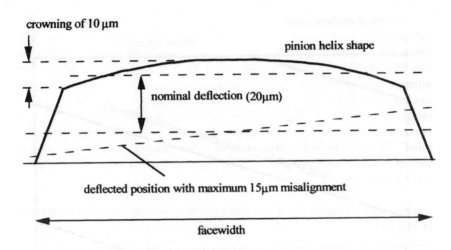

Fig 19.3 Helix matching and deflections for a compromise design.

One approach to the problem, which can be used if the "design condition" load is extremely low, is to use very heavy crowning and a perfect involute profile with merely a chamfer at the tip.

A smooth run-in is achieved thanks to the crowning, and it is permissible to dispense with conventional tip relief if loads are low since the teeth are not deflecting significantly. This type of design is quiet at low load and tolerates very high misalignments but cannot be loaded heavily as the lengths of contact line are so short. Adding tip relief to the profile allows the use of moderate loads but, as with a spur gear, we cannot get low T.E. at both design load (for which we need long relief) and low load (for which we need short relief).

In practice we do not normally have either perfect alignment or extremely low loads to allow us to use the two extreme designs described above, so compromises are necessary. Fig. 19.3 shows one possible compromise pinion helix shape where we have estimated a maximum misalignment of ±15 μm across the facewidth and expect 20 μm nominal tooth deflection. A crowning of 10 μm will keep peak deflections and loadings roughly constant provided the helix mismatch stays within 15 μm and at the ends a further end relief of 25 μm might be suitable. The wheel would then not be helix relieved at all.

Profile shape would follow normal "spur gear" rules with the choice between "long" and "short" relief according to whether best performance is required at full load or low load. Exact design of the relief is difficult because there are variations in deflections of up to 10 μm across the facewidth so design is inevitably a compromise.

The previous comments were made in relation to standard proportion 20° pressure angle gears. However, as the effect of the inevitable helix mismatch is to move the characteristics more towards those of spur gears, we can take this to the extreme and design as if they were spur gears. The ultimate spur gear design, as far as noise is concerned, is a low pressure angle tooth with an effective contact ratio of 2 (requiring a higher nominal contact ratio). The problem is slightly easier than for an actual spur gear as tip relief is not needed, just a chamfer, because a smooth run-in is achieved by the end relief. The resulting gear should be quiet at low and high load whether aligned well or not, provided that the "spur" profile has the correct long relief and a real contact ratio of 2.

The above comments apply to "rigid" gear bodies without torsional windup, without radial wheel rim deflection and without bending or distortion of overhung shafts. If any distortion or body deflection effects are occurring then their effects have to be added into the estimates. This works backwards by assuming that the loading is even across the facewidth, estimating the deflections and distortions and putting these into the calculations then re-estimating the loadings if the gear is corrected. A second iteration may be needed.

19.3 Design sensitivity

It is relatively easy, using a computer, to design a pair of gears which will be perfectly quiet under a given load. All that is then required is to make them accurately and to align the axes well in the gearbox, and we will then have a perfectly inaudible gearbox!! If only! Referring to the generation of T.E. illustrated in Fig. 19.4, it is all too clear that a dozen tolerances of 2 μm (at best) are going to have trouble fitting into a permissible T.E. of perhaps 1 μm.

The reality is that all the factors will have errors, some relatively small at 2 μm but some large at 5 to 10 μm and although elasticities will allow some averaging, there are likely to be relatively large variations.

The difficulty, and the corresponding skill, lies in having a compromise design which will be reasonably tolerant of the likely errors in a gear drive. Unlikely errors, such as having a profile on one tooth completely different from the next tooth, should not be considered but reasonable errors of profile, pitch and helix matching should be allowed for in the design.

Realistically, the only way to assess the effects is to have a computer model such as the one in section 4.5 and to vary all the tolerances by expected manufacturing errors and assess the effect both on T.E. (noise) and on peak stress loadings.

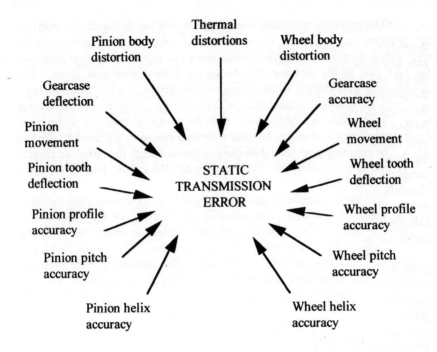

Fig 19.4 Contributors to mesh static T.E.

The effort involved is well worthwhile since it is not always obvious what effects the changes of design and manufacturing variables will have in practice, either on strength or vibration.

The danger with allowing an inexperienced designer to use a computer model is that they will take the simplistic view that whatever their design, if the computer predicts that the T.E. will be zero, then the design is "perfect." This mindset then puts all the blame for trouble on "inadequate production." It is important to educate a designer that relatively large (5 μm, 0.2 mil) profile errors and larger helix errors are inevitable and that their design must be good enough to tolerate errors, from both aspects of stressing and noise.

19.4 Buying problems

When buying-in gears, the problems fall into two groups, stress and noise, with a great difference between the degree of control and confidence in the two cases.

Currently there are few problems associated with gear strength and durability. Around the world, a few gear sets fail each year but failures are rare and invariably there have been silly mistakes made, so investigations are simple and straightforward and apportioning blame is relatively easy. Often the problem is due not to one error but to a combination of errors. As far as the buyer is concerned, specification of the drive that it should be to either the AGMA or ISO/DIN/BS specification should produce a satisfactory result. The gear manufacturers dare not produce an inadequate strength drive (because of the legal implications) so there is little to worry about. A glance at the computer printout to check that a sensible value (> 1.5) for K_β (the load intensification factor) was used and that an adequate safety factor (2) was present should be sufficient. The times this may not be adequate are if a ridiculously low diameter to length ratio was used on the pinion without helix correction or if sharp corners were left to give stress concentrations.

Noise is much more difficult. If it is the gearcase itself which is going to be the noise emitter then, as with a hydraulic power pack, specifying the total sound power emitted or specifying, say, 77 dBA at 1 m distance for a machine tool, or 60 dBA for an office device, will ensure a sufficiently quiet drive. The problem that arises in practice is that it is often not the gearbox itself that emits the sound but the main structure, as discussed in section 10.2. The only worthwhile tests are those in position in the unit and it is then all too easy to shuffle blame between gearbox and installation.

A knowledgeable customer can start by specifying a "reasonable" T.E. at each mesh in the gearbox but this requires a sophisticated investigation of the results obtained in situ with known levels of T.E. in the mesh. There are the problems of first determining a tolerable level and the associated problem that often neither the manufacturer nor the customer will yet have T.E. measuring equipment so they cannot easily check, especially since the critical value is the single flank error under load rather than under inspection conditions. Attempting to specify the necessary quality by invoking an ISO single flank quality level comes to the same thing in theory but, like the normal quality checks, takes no notice of whether it is 1/rev or 1/tooth that is important or whether both are within specification but the waveform is wrong or whether odd things happen under load so a specification may be wastefully expensive.

Overall, the depressing conclusion is that the buyer is rather in the dark for a new design and has little choice but to put their faith in a manufacturer, try the result, then if trouble occurs, panic and measure T.E. Dependent on the T.E. level the buyer can then try another manufacturer, attempt to reduce T.E levels or improve the tolerance of the installation, with economics in control as usual.

It is important, however, that initially the manufacturer is given all the relevant information since this influences the design. Apart from the obvious information about frequency of overloads or whether the drive will be idling most of its life, it is important that the designer knows what load levels are most critical for noise purposes and whether external loads are likely to distort the gearcase and affect alignments.

Units

The units used predominantly in this book are the official SI units based on kilogrammes, metres and seconds. A force of 1 Newton is defined as the force required to accelerate 1 kg at 1 m s^{-2}. The unit of work is the Joule which corresponds to the work when 1 N pushes a distance of 1 m. This is also the basic unit of all electrical work and all heat.

1 Joule per second is 1 Watt.

The standard conversions of the base units are:

$$1 \text{ lb} = 0.453592 \text{ kg}$$
$$1 \text{ inch} = 25.40000 \text{ mm}$$

From these, all the others are derived, and a particularly useful one is

$$1 \text{ lbf in}^{-2} = 6894.8 \text{ N m}^{-2}$$

so that the Modulus for steel (at 30×10^6 psi) is 210×10^9 N m^{-2}. The corresponding density is 7843 kg m^{-3}.

Stiffness conversion of 1 lbf/inch is 175.13 N m^{-1} and so a typical good machine tool stiffness of one million lbf/in is 1.75×10^8 N m^{-1}

The unit of pressure or stress, N m^{-2} is called the Pascal, written Pa, but it is rather small so a useful size for stresses is 10^6 Pa or MPa, usually written by structural engineers as N mm^{-2}. 1MPa (147 psi) is 10 bar or 10 atmospheres. For steel at 1 millistrain, the stress is 210 MPa so this is a typical working stress. In gears, working contact stresses range up to 1500 MPa (210,000 psi) for the contact stresses for a case-hardened gear.

Stiffness per unit facewidth has the same dimensions as stress and so the same conversion factor of roughly 7000 applies. This gives the "standard" tooth stiffness of 2×10^6 lbf/in/in as 1.4×10^{10} N m^{-1} m^{-1} so that a tooth 10 mm wide should have a stiffness of 1.4×10^8 N m^{-1}.

As far as general measurements, the system insists that all sizes must be quoted in millimetres on a drawing so a car may be 5683.375 long and a shim may be 0.025. Centimetres, though often used by physicists and in Europe, are illegitimate.

Also illegitimate, though not uncommon, is the kilopond, or the weight of a kilogram and 9.81 N. The acceleration due to gravity is taken as

9.81 m s^{-2}, though in practice it varies locally so it is often not possible to use dead weights accurately for force because local gravity is not known with sufficient accuracy.

It is convenient that the metric Tonne or 1000 kg has a weight of roughly 10,000 N or 10 kN which is almost exactly the same as the imperial Ton of 2240 lb.

Manufacturing accuracies are in microns (μm), roughly 0.4 tenths of a mil (thou) since 25.4 microns are 1 mil. This size of unit is ideal and is far better for quoting than "halves of tenths of a thou" for present day accuracies.

Oil viscosities start to get complicated especially as initial conversions from units such as Redwood secs are required. 1 Poise is 0.1 N s m^{-2} and 1 Stoke is 10^{-4} m^2 s^{-1}.

One great advantage is that the units of work are common to all branches of engineering so that 1 N at 1 m s^{-1} is doing 1 Watt of work and mechanical to electrical conversions are much simplified so it is easy to compare, say, energy storage in a flywheel and a capacitor.

Index